FLUID MECHANICS: ADVANCED APPLICATIONS

ROGER KINSKY

The McGraw-Hill Companies, Inc.

Sydney New York San Francisco Auckland
Bangkok Bogotá Caracas Hong Kong
Kuala Lumpur Lisbon London Madrid
Mexico City Milan New Delhi San Juan
Seoul Singapore Taipei Toronto

McGraw·Hill Australia

A Division of The McGraw·Hill Companies

Reprinted 2003, 2005, 2007, 2008, 2010
Text © 1997 Roger Kinsky
Illustrations and design © 1997 McGraw-Hill Book Company Australia Pty Limited
Additional owners of copyright are named in on-page credits.

National Library of Australia Cataloguing-in-Publication data:

Kinsky, Roger.
Fluid mechanics: advanced applications

Includes index
ISBN (10): 0 07 470442 7
ISBN (13): 9780074704424

1. Fluid mechanics. I. Title.

532

Published in Australia by
McGraw-Hill Australia Pty Limited
Level 2, 82 Waterloo Road, North Ryde, NSW 2113, Australia

Acquisitions Editor: Karolina Kocalevski
Production Editors: Ken Tate and Caroline Hunter
Permissions Editor: Natalie Muir
Designer and Illustrator: Kim Webber

Typeset in 10 on 11.5 point Times by Monoset Typesetters
Printed in Australia by Griffin Digital

Contents

Preface

This book is written essentially as a text for the engineering module *Fluid Mechanics 2*. It contains most of the topics that appeared in my earlier book *Applied Fluid Mechanics* that were not included in the 1996 book *Thermodynamics and Fluid Mechanics: An Introduction*. In addition there is a new chapter, 'Channel flow', which appears as Chapter 4.

Most chapters have a greater depth of content than the earlier book *Applied Fluid Mechanics*. Also there are more illustrations and worked examples, as well as the inclusion in each chapter of objectives, a summary and self test problems.

The essential features of and changes in each chapter are as follows:

Chapter 1
A short revision of fluid properties and flow principles is given. Non-circular pipes are treated, as well as circular ones.

Chapter 2
The Moody formula is given as an alternative to the Moody diagram as a method of finding the friction factor.

Chapter 3
The section on pipe flow has been expanded considerably and treats a wide variety of free-flow problems, including those involving series and parallel pipes. Often an iterative solution is required.

Chapter 4
As well as traditional treatment of channel flow using the Chezy and Manning formulas, the solution using the Darcy equation is given first as a logical extension of pipe flow problems using this equation.

Chapter 5
As well as treatment of rotodynamic pumps, this chapter includes a description of most of the popular types of positive-displacement pumps. The performance of both types of pump is treated, as well as the effect on performance of a speed change.

Chapter 6
This chapter deals with pump selection and pump performance in a system, with emphasis on rotodynamic types, but positive-displacement types are also treated. *NPSHA* is treated using the Australian Pump Manufacturers' Association method based on the total head (including velocity head) at the pump inlet and not just the pressure head.

I have attempted to make the book as practically oriented and 'user friendly' as possible for both learners and teachers. There are many illustrations and worked examples as well as solved self-test problems. These self-test problems can be used in a self-paced learning environment or as class exercises. There are numerous appendixes containing important data and diagrams as well as some formula derivations.

All important formulas are numbered and boxed and the meaning of the symbols in each formula and the correct units for formula substitution are stated.

Each chapter commences with a list of objectives and an introduction, and concludes with a summary and a comprehensive set of problems requiring both descriptive and analytical solution. There are sufficient problems to enable them to be used as a 'question bank' for assessment purposes.

The answers to the analytical problems are given at the end of each problem. A manual containing the solutions to all problems in the book is available and will be issued free of charge to all adopters of the title as a textbook. Copies of the manual can be obtained from the publisher, McGraw-Hill.

Thanks go to George Stratford (Meadowbank TAFE) for producing a CAD drawing of the centrifugal pump impeller that was used in Figure 5.21, and to the following reviewers for their valuable feedback and suggestions: Carl Ackland, Southern Sydney Institute of TAFE, NSW; John Arnold, Regency Institute of TAFE, SA; Bob Barnes, Moreton Institute of TAFE, QLD; Dallas Elvery, Northern Territory University, NT; Vic Ilic, University of Western Sydney, NSW; Ian Leggo, Open Training and Education Network, NSW; Val Newfield, Illawarra Institute of Technology, NSW; Bill Reynolds, Launceston Institute of TAFE, TAS; Joe Turco, Open Training and Education Network, NSW; Roel Wasterval, Western Metropolitan College of TAFE, VIC. I am also indebted to Thompsons, Kelly and Lewis for permission to reproduce copyright material: Figure 5.31 and Appendix 12, 'Centrifugal pump performance curves', and Figure 6.5 and Appendix 14, 'Range chart'.

While every attempt has been made to produce a book free of errors, the practicality is that some errors will no doubt appear. I would be most grateful to have these brought to my attention. I would also be pleased to hear from any users of the book who might have any comments or suggestions for improvement.

ROGER KINSKY

Sydney, January 1997

CHAPTER 1

Reynolds number and flow regime

Objectives

On completion of this chapter you should be able to:

- describe and sketch the flow regimes of laminar and turbulent flow;
- describe transition (mixed) flow;
- describe the three main factors that influence the flow regime and recognise that these can be expressed in the form of a dimensionless number—the Reynolds number;
- calculate the Reynolds number for a flowing fluid;
- calculate the equivalent diameter, and use this to calculate the Reynolds number for fluid flowing in non-circular or partly-full pipes;
- state the relationship between Reynolds number and flow regime;
- define the upper and lower critical Reynolds number;
- describe and sketch typical velocity profiles for laminar and turbulent flow.

Introduction

In *Thermodynamics and Fluid Mechanics: An Introduction*, the basic properties of fluids and the basic principles of fluid flow are outlined. This chapter extends this basic knowledge into the investigation of flow regimes and their relationship to the Reynolds number. However, before doing so a short revision of fluid properties and flow principles is given.

1.1 *FLUID PROPERTIES*

The important fluid properties used throughout this book are density (also expressed as relative density or specific volume), pressure, temperature, and viscosity.

Density

Density ρ is the mass per unit volume and the units are kg/m^3. That is:

$$\rho = \frac{m}{V}$$

Density is used mainly with solids and liquids, and is a measure of the size of the molecules and how closely they are spaced. The density of fresh water (at 0°C and 101.325 kPa abs.) is 1000 kg/m^3 or 1 kg/L.

Specific volume

Specific volume (v) is volume per unit mass and therefore it is the reciprocal of density. That is:

$$v = \frac{V}{m}$$

Units of specific volume are m^3/kg. Specific volume is used with gases (although the density of a gas may also be stated). For air at 20°C, the specific volume is about 0.85 m^3/kg and the density is about 1.2 kg/m^3.

Relative density

Relative density (*RD*) is the density of the substance, divided by the density of fresh water taken as 1000 kg/m^3. Relative density has no units. If a homogeneous substance that does not mix with water has a relative density less than 1 (one), the substance will rise to the surface and float in fresh water. Conversely, if the relative density is greater than 1, the substance will sink in fresh water. For example, kerosene and oils with *RD* less than 1 float on the surface of water, whereas mercury, with *RD* about 13.6, sinks.

Pressure

Pressure p is force per unit area. Units are Pa (pascals) where 1 Pa = 1 N/m^2. With high pressure fluids the pressure is usually given in kPa or MPa.

Atmospheric pressure (the pressure of air at sea level) is about 101.3 kPa and unless otherwise stated will be assumed to have this value throughout this book. Most pressure gauges measure pressure above or below atmospheric, and pressure measured or stated this way is known as *gauge pressure*. Pressure may also be given as *absolute pressure* which is pressure measured above absolute zero pressure. It is important to be clear whether a given pressure is absolute or gauge. In most cases in fluid mechanics, gauge pressures are used, so in this book, *assume that all given pressures are gauge pressures unless otherwise stated*. Absolute pressure cannot be negative, but gauge pressure can have a negative value. Negative gauge pressure is also known as *vacuum*, that is, pressure below atmospheric.

Effect of pressure on density

The effect of pressure on density (or specific volume) of liquids is very different to the effect on gases.

In liquids, little change in density occurs with large changes in pressure. For example, the density of sea water at the surface of the ocean is only slightly less than the density at the ocean floor, even though the pressure at the ocean floor can be hundreds of times greater than the pressure at the surface.

In gases, a significant change in density (or specific volume) occurs with changes in pressure. The higher the pressure, the lesser is the specific volume and the greater the density. For example, air pressure at sea level is considerably higher than air pressure at the top of a high mountain. Consequently, air density at sea level is considerably higher than air density on the top of the mountain.[1]

The ability of gases to reduce volume with increasing pressure is known as compressibility.

Relationship between fluid pressure and depth

For a liquid (or other homogeneous substance of constant density), the relationship between pressure due to self-weight and depth is given by the important formula:

$$p = \rho g h$$

where p = increase in pressure (Pa) due to self-weight
         ρ = density of the liquid (kg/m^3)
         g – gravitational constant (9.81 N/kg or 9.81 m/s^2)
         h = vertical depth (m) below the free surface

Note   In most cases, atmospheric pressure acts on the liquid surface, which is therefore at zero gauge pressure. In such cases the increase in pressure given by this formula is equal to the gauge pressure at depth h.

Temperature

Simply stated, temperature T is the degree of hotness or coldness of a substance. More precisely, temperature is a measure of the average linear kinetic energy of the molecules of a substance. That is, with increasing temperature the molecules of a substance move more rapidly. In solids and liquids, molecular motion is primarily vibrational (back and forth) whereas in gases, molecular motion is primarily translational.

Temperature has the unit kelvin (K) which is temperature measured above absolute zero temperature. At absolute zero the molecules of a substance do not move at all. Temperature can also be stated in Celsius units, which is the temperature measured above the freezing point of fresh water taken as the zero point. This temperature is about 273 K, so:

$$T \, (K) = T \, (^{\circ}C) + 273$$

Effect of temperature on density

The effect of temperature on the density (or specific volume) of liquids is very different to the effect on gases. In liquids, little change occurs. For example, referring to Appendix 13 it is seen that in the temperature range 0–20°C the change in the density of water is about 0.2% and therefore negligible. If the temperature is raised from 0 to 100°C the decrease in density is only about 4%. However, if the same temperature increase occurs in a mass of air (at constant atmospheric pressure) there will be a decrease in density of about 37%.

---

[1] This is the reason why aircraft fly at high altitudes on long trips, because at high altitudes the density and air resistance are much less.

The relationship between temperature, pressure, volume or density for a constant mass of gas (considered perfect or ideal) is given by the following formulas:

$$\frac{p_1 V_1}{T_1} = \frac{p_2 V_2}{T_2}, \quad pV = mRT, \quad p = \rho RT, \text{ and } R = \frac{8314}{M}$$

where
p = pressure in Pa (abs.)
V = volume in m^3
T = temperature in K
m = mass of the gas in kg
ρ = density of the gas in kg/m^3
R = gas constant (J/kg K)
M = relative molecular mass of the gas (no units)

Note   Because the density of water varies so little under normal circumstances in fluid mechanics, in this book the density of fresh water will be taken as 1000 kg/m^3, unless an alternative value is given, or the water is at an elevated temperature (above about 20°C).

Viscosity

An important distinction between solids and fluids is that fluids can flow, whereas solids cannot. However, a flowing fluid has a resistance to motion very similar to the frictional resistance to motion exhibited by solids. In fluids, the resistance to flow is known as *dynamic viscosity*, *absolute viscosity*, or simply *viscosity*.

In this book, the Greek symbol μ (pronounced 'mew') is used for dynamic viscosity. Units are pascal seconds, Pas.

Thick liquids such as honey, or thick oil, have a high viscosity, whereas thin liquids such as water and alcohol have a lower viscosity than thick liquids. As would be expected, gases have a lower viscosity than liquids. A graph showing the viscosity of some common fluids is given in Appendix 3. This graph is valid for a reasonable range of pressures above or below atmospheric. Also the viscosity of water in the range 0–100°C is tabulated in Appendix 13. Note that viscosity varies considerably with temperature and the viscosity of liquids decreases with temperature, whereas the opposite is true for gases.

Kinematic viscosity

Frequently in fluid mechanics formulas the ratio $\dfrac{\mu}{\rho}$ appears, that is, the dynamic viscosity divided by the density of the fluid. This ratio is called kinematic viscosity and in this book the Greek symbol v (pronounced 'new') is used for it. Units of kinematic viscosity v are m^2/s.

The relationship between kinematic and dynamic viscosity is given by the formula:

$$v = \frac{\mu}{\rho}$$

Note   In a problem, if it is not stated whether the viscosity is dynamic or kinematic, this can be determined from the units.

1.2 *PRINCIPLES OF FLUID FLOW*

The important basic principles of fluid flow are now revised.

Steady flow

Steady flow occurs when the fluid flow characteristics (such as velocity or pressure) do not change with time. For example, when a water tap in a house is opened, flow out of the tap is initially unsteady because the flow rate is changing (increasing from zero to some value). If the tap is then left in the open (or partly open) position, steady flow becomes established (provided the water supply pressure remains constant).

This book deals only with fluids in steady flow.

Velocity

The velocity of a fluid in flow (v) is the *average* or *mean velocity*. When fluid flows in a pipe or duct, not all molecules across any section move in the same direction and with the same speed. This can be seen clearly by observing the flow of water in a river or stream as shown in Figure 1.1.

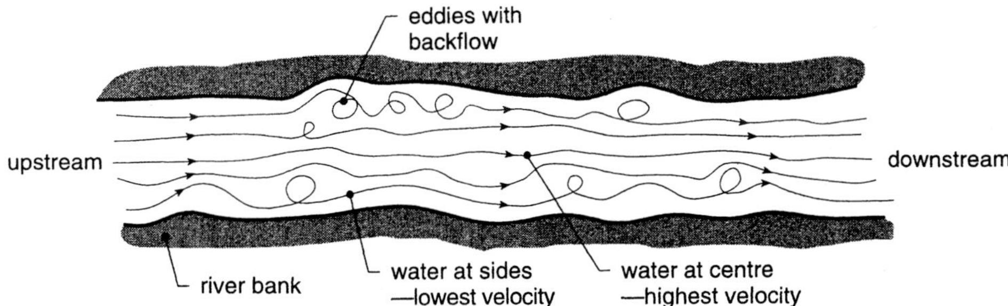

Fig. 1.1 *Water flow in a river*

The water at the centre flows faster than the water near the banks. Also, at the banks there are often eddies, where some water flows with a circular motion. This means that some water in the eddy is actually flowing backward relative to the main stream flow.

Consider a river where the flow rate in the downstream direction is 20 m³/s . At some section in the river the cross-sectional area is 10 m². The water velocity at this section is 2 m/s because this is the average flow velocity there.

Volume flow rate

Volume flow rate $\dot{V}$ is the volume of fluid flowing per unit time (usually per second). The volume flow rate of fluid with velocity v flowing through a section of cross-sectional area A is:

$$\dot{V} = vA \quad (\text{m}^3/\text{s} = \text{m/s} \times \text{m}^2)$$

The dot above the V is used throughout this book to mean *per unit time*. Base or SI units of volume flow rate are m³/s but L/s units are also often used (1000 L/s = 1 m³/s).

Mass flow rate

Mass flow rate $\dot{m}$ is the mass of fluid flowing per unit time (usually per second). Mass flow rate is obtained by multiplying the volume flow rate by the density of the fluid. That is:

$$\dot{m} = \dot{V}\rho = vA\rho \quad (\text{kg/s} = \text{m/s} \times \text{m}^2 \times \text{kg/m}^3)$$

Again, the dot above the m denotes per unit time. The unit of mass flow rate is kg/s.

Continuity equation

The continuity equation is the conservation of mass law applied to fluid flow through a system. Under steady flow conditions, the mass flow rate into a system is equal to the mass flow rate out of the system. This is true *regardless of the nature of the fluid* and applies to gases and vapours as well as liquids.

Mathematically, the continuity equation is:

$$\Sigma \dot{m}_{\text{in}} = \Sigma \dot{m}_{\text{out}} = \text{constant}$$

That is, the sum of the mass flow rates into the system is equal to the sum of the mass flow rates out of the system.

Denoting the inlet as ① and the outlet as ② (as shown in Figure 1.2), the continuity equation can also be written:

$$v_1 A_1 \rho_1 = v_2 A_2 \rho_2$$

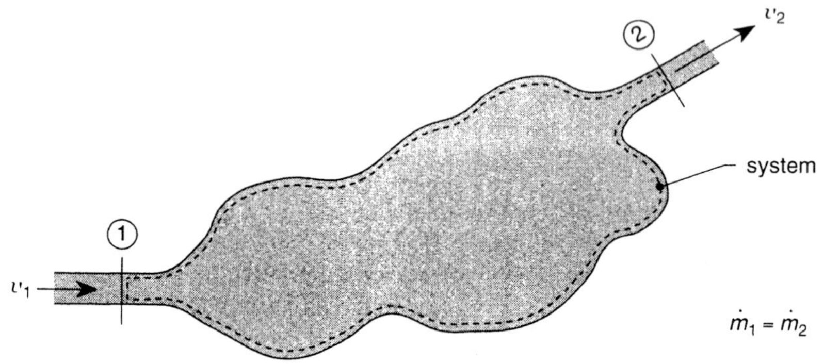

Fig. 1.2   *Fluid flow through a system*

For *liquids* or gases where the density is approximately constant (small change in pressure or temperature), the continuity equation may be written in terms of volume flow rates :

$$\Sigma \dot{V}_{\text{in}} = \Sigma \dot{V}_{\text{out}} = \text{constant}$$

or in the alternative form:

$$v_1 A_1 = v_2 A_2$$

As a revision exercise, Self-test problem 1.1 should now be attempted.

Self-test problem 1.1

Liquid with relative density 0.8 flows at a rate of 1.2 kg/s through the pipe shown in Figure 1.3.

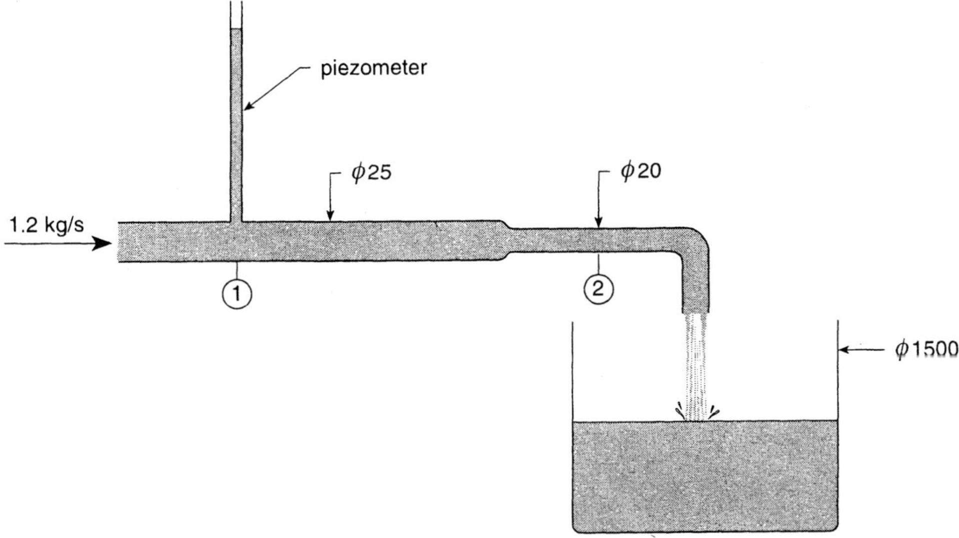

Fig. 1.3

The pressure at ① is 110 kPa (abs.) at the centre of the pipe.
Determine:
(a) the velocity at ①
(b) the velocity at ②
(c) the height in mm shown by the piezometer at ① (to the centre of the pipe)
(d) the time (in hours) for 6 m³ of liquid to flow into the tank
(e) the height of liquid in the tank when it contains 6 m³ of liquid
(f) the pressure (gauge and absolute) at the base of the tank when it contains 6 m³ of liquid.

1.3　FLOW REGIMES

There are two essentially different flow regimes that can occur with fluids and these are known as *laminar* and *turbulent*. These regimes, and the circumstances causing them, were first investigated comprehensively by Osborne Reynolds in 1883 using the apparatus illustrated in Figure 1.4 on page 8.

The flow rate of water through the transparent tube could be adjusted by the valve at the end of the tube. The appearance of the water flowing through the tube could be seen by injecting dye into the water just before it entered the tube.

Laminar flow

Sometimes filaments of dye streamed along in the water without mixing through the water. Reynolds concluded that the water was flowing in layers or laminas, so he termed this *laminar flow*. As previously stated, when a fluid flows in a pipe or tube the fluid at

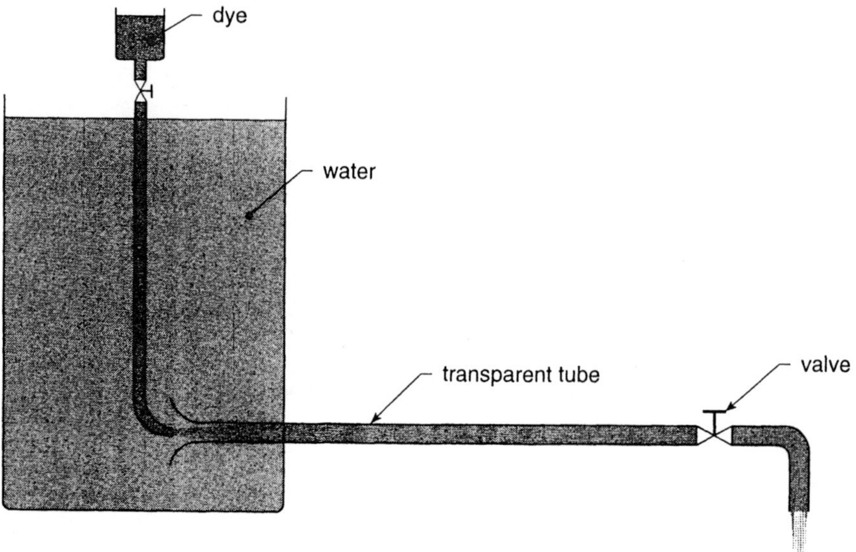

Fig. 1.4   *Reynolds apparatus*

the centre is moving faster than the fluid at the outsides. At the wall surface a thin layer of fluid clings to the wall.[2] Away from the wall the velocity increases, and reaches a maximum value at the centre of the pipe. Therefore there is a shearing action through the fluid, just as happens when a pack of cards is held at top and bottom and the centre cards are pushed through the pack. Laminar flow and the card analogy to laminar flow are illustrated in Figure 1.5.

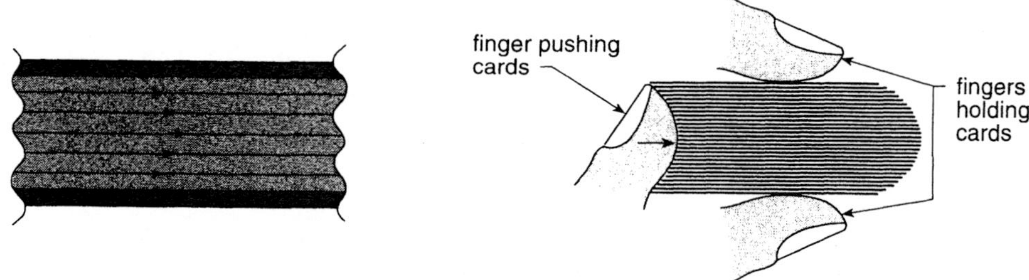

Fig. 1.5   *Laminar flow and the card analogy*

Another way of visualising laminar flow is to imagine the particles of fluid moving along in lanes like cars on a freeway. If there is no lane-changing, the motion of the cars resembles particles of fluid in laminar flow.

---

[2] Observe that a dusty car remains dusty even when driven at high velocity. The dust on the surface is in the stationary fluid layer and does not blow off.

Turbulent flow

Sometimes the filaments of dye quickly intermingled and diffused throughout the water. Reynolds concluded that mixing was taking place, that is, the water was no longer flowing in layers or laminas but moving around with a swirling motion. Reynolds termed this *turbulent flow*. Turbulent flow is illustrated in Figure 1.6.

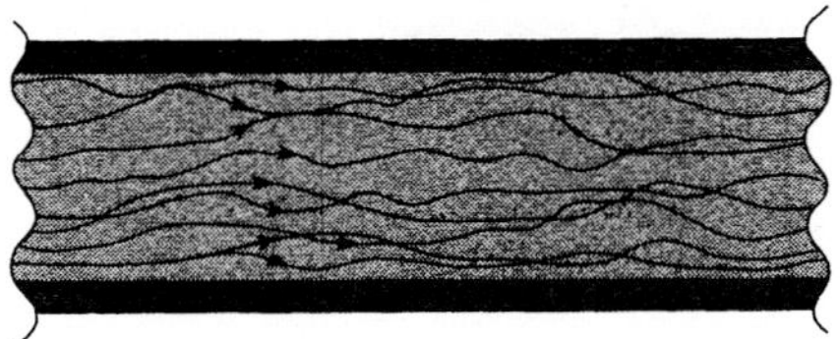

Fig. 1.6   *Turbulent flow*

In turbulent flow the fluid particles move back and forth across the fluid stream, as well as along it. Turbulent flow can be visualised as the particles of fluid moving along like cars on a freeway where constant lane-changing occurs.

Turbulent flow is by far the most common type of flow encountered in gases and liquids flowing in ducts and pipes. Turbulence can occur when a fluid flows along a straight length of duct or pipe, but it can also be induced by *turbulence promoters* (Figure 1.7). Fluid machinery such as fans, pumps, propellers and stirrers induce turbulence due to the swirling action and pulsations imparted to the fluid. Roughness or scale at the wall of a pipe or duct tends to induce turbulence; so too do valves or fittings (particularly when these have sharp projections or corners). As a general rule, any obstacles in the flow path act as turbulence promoters, as when outside air flows over trees or buildings, or when inside air flows over furniture and fixtures. In air-flow ducts, mixing boxes are sometimes used to promote turbulence and ensure thorough mixing of air streams. Turbulence may also be induced by external vibration, or movement of a fluid system, for example in a factory where operating machinery causes vibration.

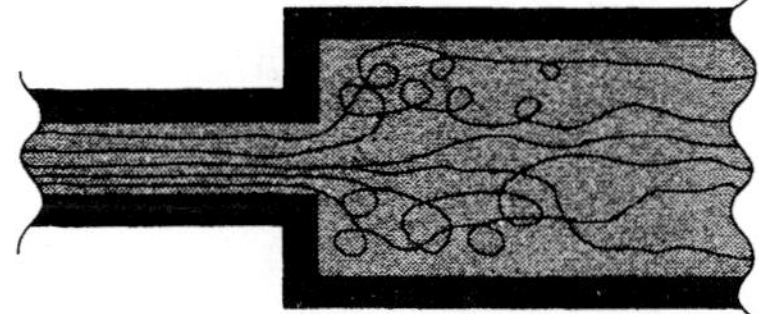

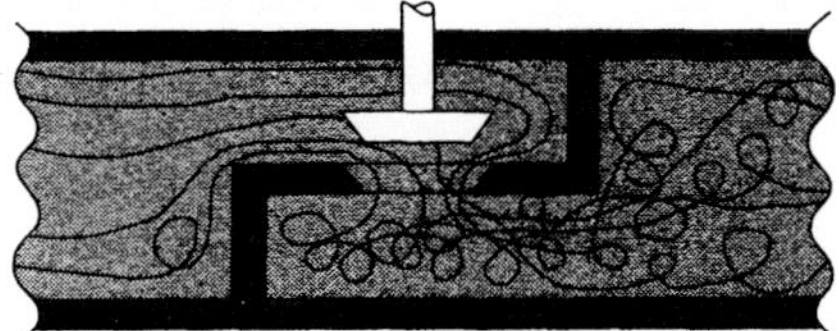

(a) Sudden enlargement

(b) Globe valve

Fig. 1.7   *Two examples of turbulence promoters*

Transition or mixed flow

Often, both laminar and turbulent flow regimes occur simultaneously. For example, in pipe flow, the fluid next to the wall is in a laminar sublayer (or boundary layer) whereas the fluid farther out from the wall can be in turbulent flow. Also, flow can be initially laminar, and then turbulence can occur downstream because of pipe roughness or turbulence promoters. Conversely, turbulence induced by stirring, or turbulence promoters in the fluid path, can damp out, and the flow can change to laminar downstream. In either case there is a region where the flow is changing from laminar to turbulent, or vice versa. The flow in this region is usually known as *transition flow* or *mixed flow* because both laminar and turbulent flow are present at the same time.

1.4 *REYNOLDS NUMBER*

Having identified the various flow regimes, Reynolds attempted to identify the factors (or variables) that influenced the flow regime and the relationship between them. That is, to answer the question: 'If fluid flows through a straight duct or pipe, will the flow be laminar or turbulent?' He found that three main factors influenced the flow regime:

- fluid velocity: the higher the velocity, the greater is the tendency for turbulence;
- fluid viscosity: the lower the viscosity, the greater is the tendency for turbulence;
- pipe diameter: the larger the pipe diameter, the greater is the tendency for turbulence.

Reynolds was able to express the relationship between these variables by a single number (now known as the Reynolds number) given by:

$$Re = \frac{\upsilon d}{\nu} = \frac{\upsilon d \rho}{\mu}$$

Reynolds number (1.1)

where υ = velocity of the fluid in m/s
d = diameter of the pipe or tube in m
ρ = density of the fluid in kg/m^3
ν = kinematic viscosity of the fluid in m^2/s
μ = dynamic viscosity of the fluid in Pas

The smaller the Reynolds number, the greater is the likelihood of laminar flow, and the greater the Reynolds number, the greater is the likelihood of turbulent flow. That is, laminar flow is likely only with relatively high viscosity fluids flowing with relatively low velocity through relatively small bore pipes or tubes.

Reynolds number is a *dimensionless* number, that is it has no units (when a consistent system of units is used). This is outlined in Self-test problem 1.2:

Self-test problem 1.2

Show that Reynolds number is dimensionless, using both forms of Formula 1.1.

Example 1.1

Calculate the Reynolds number when water at 50°C flows at a rate of 3 L/s through a pipe of diameter 30 mm.

Solution

From Appendix 13 for water at 50°C: with viscosity = 0.55 × 10^{-3} Pas, and density = 988 kg/m^3

$$\upsilon = \frac{\dot{V}}{A} = \frac{3 \times 10^{-3}}{A(0.03)} = 4.244 \text{ m/s}$$

The viscosity is the dynamic viscosity μ, so the Reynolds number formula that should be used is:

$$Re = \frac{\upsilon d \rho}{\mu} \quad \left(\frac{\text{m s}^{-1} \times \text{m} \times \text{kg m}^{-3}}{\text{Pas}} \right)$$

Substituting:

$$Re = \frac{4.244 \times 0.03 \times 988}{0.55 \times 10^{-3}}$$

$$= 228\,720$$

$$= \mathbf{229} \times \mathbf{10^3} \text{ (to 3 significant figures)}$$

Equivalent diameter

When a fluid flows in a pipe or duct other than a circular one, the Reynolds number can be calculated using the equivalent diameter defined as:

$$\boxed{d_e = \frac{4A}{P}}$$

Equivalent diameter (1.2)

where A = cross-sectional area of the fluid in flow
   P = cross-sectional wetted perimeter

The equivalent diameter d_e for a non-circular pipe may then be used in place of diameter d for a circular pipe in Formula 1.1 in order to calculate the Reynolds number.

Notes

- The use of equivalent diameter does not produce exact results in fluid flow calculations. Some error can occur and this increases as the section diverges from a circular one. For example, when using the equivalent diameter for a very flattened duct with width much greater than height, a greater error occurs than is the case with a square or slightly oval duct section.
- When liquid flows in a pipe or duct that is not running full, the cross-sectional area of fluid flow and the wetted perimeter are not the same as when the liquid completely fills the pipe or duct. This situation does not normally arise with gases, which almost always completely fill the duct or pipe conveying them.
- For a circular section with fluid running full, $A = \dfrac{\pi d^2}{4}$ and $P = \pi d$

  so $d_e = \dfrac{4\pi d^2}{4} \bigg/ \pi d = d$ (which is to be expected).
- For a square section of side d with fluid running full, $A = d^2$ and $P = 4d$

  so $d_e = \dfrac{4d^2}{4d} = d$ (which is the same as a circular section inscribed in the square).

- The ratio $\dfrac{A}{P}$ is also known as the hydraulic radius R (not to be confused with the radius of a circle).

Example 1.2

A sloped rectangular channel of width 500 mm carries water to a depth of 300 mm. The water flows with velocity 1.2 m/s. Calculate the Reynolds number, given that the viscosity of the water is 1.14×10^{-6} m^2/s.

Solution

The wetted area is $0.5 \times 0.3 = 0.15$ m^2
The wetted perimeter is $0.5 + 0.3 + 0.3 = 1.1$ m
The equivalent diameter d_e is $4 \times 0.15/1.1 = 0.5455$ m
In this case kinematic viscosity is given (in m^2/s) so the Reynolds number is given by:

$$Re = \frac{vd_e}{v} = \frac{1.2 \times 0.5455}{1.14 \times 10^{-6}} = 574 \times 10^3$$

 ## Self-test problem 1.3

Air flows with velocity 3 m/s in a rectangular duct of cross-section 300×200 mm. Calculate the Reynolds number, given that the viscosity of the air is 18.5×10^{-6} Pas and the density is 1.18 kg/m^3.

1.5 CRITICAL REYNOLDS NUMBER

As was stated previously, the smaller the Reynolds number the greater is the tendency for laminar flow, and the larger the Reynolds number, the greater is the tendency for turbulent flow. The question remains: 'If no turbulence promoters are present, at what values of Reynolds number do laminar or turbulent flow occur in pipes?'

Reynolds found that for values of the Reynolds number below 2000, laminar flow will occur. Even if the fluid flow is originally turbulent (perhaps as a result of pumping or stirring) the turbulence will dampen out downstream and laminar flow will become established if the Reynolds number is less than 2000. This value is known as the **lower critical Reynolds number** and is the value of the Reynolds number below which laminar flow occurs.

The value of the Reynolds number above which turbulence occurs is known as the **upper critical Reynolds number** and this value is not as well defined as the lower critical Reynolds number. Reynolds showed that laminar flow could occur with Reynolds numbers as high as 40 000 if special precautions were taken to settle the water in the tank for a long time, and to prevent any vibration or bumping of the water. These special conditions do not generally occur in industry so for most fluid applications in engineering the upper critical Reynolds number can be taken as 4000.

For the range of Reynolds numbers between 2000 and 4000 a region of uncertainty exists where any of the flow regimes can occur.

These conclusions may be summarised as in Table 1.1:

Table 1.1   *Reynolds number and flow regime*

Reynolds number	Flow regime
Below 2000	laminar
2000 – 4000	uncertain (critical region)
Above 4000	turbulence is most likely

Note   In this book, it is assumed in all cases that if the Reynolds number is greater than 4000, flow is turbulent, and if less than 2000, flow is laminar.

Example 1.3

Oil with viscosity 0.2×10^{-3} m^2/s is stirred in a tank and then pumped through a pipe of diameter 100 mm. What is the maximum velocity of the oil in the pipe if, downstream of the tank, turbulence damps out and laminar flow becomes established?

Solution

Because kinematic viscosity is given, use the Reynolds number formula:

$$Re = \frac{\upsilon d}{\nu}$$

The lower critical Reynolds number is 2000, which is the value of the Reynolds number below which laminar flow occurs,

$$\therefore 2000 = \frac{\upsilon \times 0.1}{0.2 \times 10^{-3}}$$

$$\therefore \upsilon = \mathbf{4\ m/s}$$

Self-test problem 1.4

Water at 20°C flows through a small-bore tube with velocity 0.5 m/s. Determine the maximum diameter of the tube for laminar flow to be certain. Use Appendix 13 for the density and viscosity of water at the given temperature.

1.6 VELOCITY PROFILES

As stated previously, when a fluid flows, there is not a constant velocity across a section. For fluid flowing in a pipe or tube it has been found that there is a characteristic velocity profile associated with laminar and turbulent flow. See Figure 1.8.

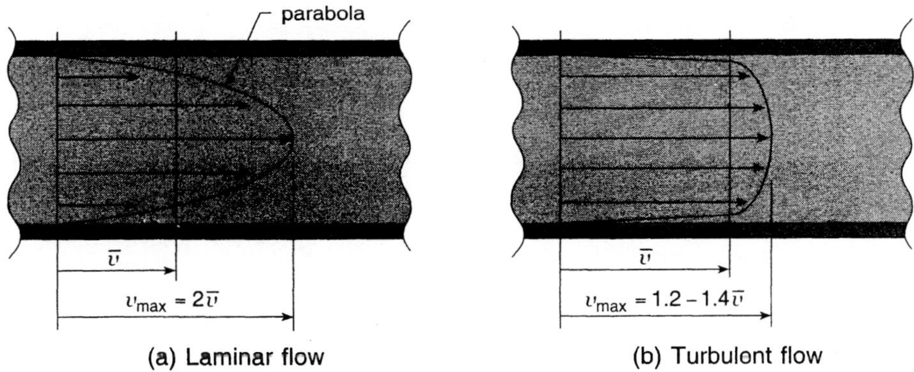

Fig. 1.8 *Velocity profiles for laminar and turbulent flow*

For laminar flow, the velocity profile is a parabola. Because the volume of a paraboloid of revolution is one-half that of the surrounding cylinder, the mean velocity is one-half the maximum velocity (the same volume of fluid passes through). That is, the maximum velocity is twice the mean velocity. The velocity profile for laminar flow is essentially independent of the roughness of the pipe, that is, the same velocity profile occurs with both rough and smooth pipes.

When turbulence occurs. the velocity profile does not have a shape that can be expressed as a single mathematical function. Rather the shape depends on the roughness of the pipe and the Reynolds number. At the walls there is a steep velocity gradient in the region of the boundary layer (laminar sublayer). Although this layer is usually very thin (of the order of only several hundreths of a millimetre) it is an important region because in it the velocity changes rapidly from zero to a value close to the mean velocity. Outside the boundary layer the velocity profile is a relatively flat curve with relatively small change in velocity. Consequently when turbulence occurs, the maximum velocity is closer to the average velocity than when flow is laminar. Usually, when flow is turbulent the maximum velocity is about 1.2–1.4 times the mean velocity.

Example 1.4

Oil with relative density 0.9 and viscosity 0.048 Pas flows at a rate of 10 L/s in a pipe of diameter 150 mm. Determine the mean and maximum velocity of the oil in the pipe.

Solution

The mean or average velocity is given by :

$$
\begin{aligned}
v &= \frac{\dot{V}}{A} \\
&= \frac{10 \times 10^{-3}}{A(0.15)} \\
&= \textbf{0.566 m/s} \\
Re &= \frac{v d \rho}{\mu} \\
&= \frac{0.566 \times 0.15 \times 900}{0.048} \\
&= 1592
\end{aligned}
$$

Since 1592 < 2000. flow is laminar. and hence the maximum velocity is twice the mean velocity, that is: $2 \times 0.566 = \textbf{1.13 m/s}$.

Summary

Some of the most important properties of fluids are density, relative density, specific volume (gases), pressure, temperature, and viscosity. For a study of fluid mechanics it is necessary to have a clear understanding of these properties and their units.

When a fluid flows through a system, flow is steady when the fluid velocity or pressure at any position does not change with time. Unless stated otherwise, the velocity of a fluid is taken to be the mean or average velocity, calculated as the volumetric flow rate at any section in the system divided by the cross-sectional area of the fluid. The mass flow rate is the volumetric flow rate multiplied by the density of the fluid.

For steady flow the continuity equation applies, so the mass flow rate into the system is equal to the mass flow rate out of the system. If the fluid has constant density (for example a liquid) the volume flow rate into the system is also equal to the volume flow rate out of the system.

Two different flow regimes are possible: laminar or turbulent flow. When the flow is laminar the fluid flows in layers and there is no intermingling of the layers. When turbulence occurs, particles of fluid move in a lateral direction as well as in an axial direction and also usually have a swirling motion. A predominantly turbulent flow regime is the most usual one that occurs.

Often both laminar and turbulent flow are present at the same time, or the flow is changing from laminar to turbulent, or vice versa. The flow in this region is known as transition or mixed flow.

When fluid flows in a pipe or duct, three factors have a great influence on the likely flow regime: the fluid velocity, fluid viscosity and diameter of the pipe. These factors or variables can be combined in a single number known as the Reynolds number. It can be calculated using either of two formulas. The one to use depends on whether dynamic or kinematic viscosity has been given, that is, whether the units of viscosity are Pas or m^2/s.

The relationship between the Reynolds number and the flow regime can be stated as follows: *The smaller the Reynolds number, the greater is the likelihood of laminar flow, and the greater the Reynolds number, the greater is the likelihood of turbulent flow.* The upper and lower limits are known as the upper and lower critical Reynolds numbers, and have values of 4000 and 2000 respectively (for fluid flow in pipes or ducts). Between these values the flow regime is uncertain.

If the pipe or duct is not circular, or if the fluid is a liquid and does not completely fill the pipe or duct, the Reynolds number can be determined using an equivalent diameter calculated as four times the cross-sectional area, divided by the wetted perimeter.

When a fluid is in laminar flow in a pipe the velocity profile is a parabola and the maximum velocity is twice the average or mean velocity. When turbulence occurs the velocity profile is a compound curve showing a very rapid change in velocity through the boundary layer at the wall, and then a relatively flat curve through to the centre where maximum velocity occurs. Then the maximum velocity is usually about 1.2–1.4 times the average velocity.

 # Problems

Note   For these problems assume the density of water is 1000 kg/m^3.

1.1 **(a)** With the aid of neat sketches, describe briefly the difference between laminar and turbulent flow.

(b) Describe briefly what is meant by transition or mixed flow.

(c) State at least five factors that can induce turbulence when a fluid flows.

1.2 **(a)** What are the three main factors or variables that influence the likely flow regime when a fluid flows in a pipe or duct? Also describe the effect that these factors have on the tendency for turbulence to occur.

(b) Write the two formulas for Reynolds number, explain when each form should be used and briefly explain each term in the formula, stating the SI units.

(c) What are the units of Reynolds number?

1.3 **(a)** If fluid flows in a pipe or duct other than a circular one, how may the Reynolds number be calculated?

(b) Define lower and upper critical Reynolds number, and state their generally accepted values.

(c) What flow regime is likely if the Reynolds number falls between the lower and upper critical values?

1.4 **(a)** With the aid of a neat sketch, briefly discuss Reynolds' apparatus and experiments on the flow regimes encountered when fluids flow in a pipe.
(b) If special conditions are present, laminar flow can occur at values of the Reynolds number well above 4000. Briefly describe these conditions.
(c) Draw up a table showing the relationship between Reynolds number and flow regime.

1.5 **(a)** With the aid of a neat sketch, describe the velocity profiles that occur with laminar and turbulent flow. On the sketch, show mean and maximum velocities in each case.
(b) What effect does pipe roughness have on the velocity profile when flow is:
  (i) laminar?
  (ii) turbulent?

1.6 Liquid with relative density 1.12 and viscosity 1.65×10^{-3} Pas is pumped at a rate of 2.5 L/s. Determine:
(a) the diameter of the inlet and outlet pipes to the pump (to the nearest mm) if the velocities in the inlet and outlet pipes are to be 1.5 m/s and 3 m/s respectively:
(b) the Reynolds number in the inlet and outlet pipes if the velocities in the inlet and outlet pipes are 1.5 m/s and 3 m/s respectively and their diameter is as determined in (a).
  (a) 46 mm, 33 mm (b) 46.8×10^3, 67.2×10^3

1.7 Methane gas (relative molecular mass 16) flows at a rate of 1.5 kg/s in a circular duct of constant diameter 400 mm fitted with a cooling coil. The temperature of the gas is 20°C upstream of the cooling coil and the pressure is 10 kPa (gauge). The cooling coil cools the methane to –20°C at constant pressure. Determine the velocity of the gas:
(a) upstream of the cooling coil;
(b) downstream of the cooling coil.
  (a) 16.3 m/s (b) 14.1 m/s

1.8 A diesel engine consumes fuel (relative density 0.84) at a rate of 25.5 L/h. The average air/fuel ratio is 17:1 (by mass). The exhaust gas just before leaving the exhaust pipe is at atmospheric pressure and a temperature of 350°C. The gas constant for the exhaust gas can be taken to be the same as for air, that is, 287 J/kgK. Determine:
(a) the mass flow rate of exhaust gas in kg/s;
(b) the minimum necessary diameter of the exhaust pipe if the velocity of the gas in the pipe just before exit (when pressure is atmospheric) is not to exceed 20 m/s.
  (a) 0.107 kg/s (b) 110 mm

1.9 Water flows through an inlet pipe of diameter 50 mm with constant velocity 2 m/s into a storage tank of diameter 1500 mm mounted with the axis vertical. An outlet pipe of diameter 30 mm allows water to flow out of the tank at a constant rate of 2.5 L/s. The viscosity of the water is 1.2×10^{-6} m²/s. Determine:
(a) the time taken for the level in the tank to rise by 1 m;
(b) the Reynolds number and flow regime in the inlet pipe;
(c) the Reynolds number and flow regime in the outlet pipe;
(d) the Reynolds number and flow regime in the tank.
  (a) 20.6 minutes (b) 83.3×10^3, turbulent (c) 88.4×10^3, turbulent
  (d) 1009, laminar

1.10 An SAE 30 lubricating oil has a relative density of 0.967 at 20°C. The oil flows in a pipe of diameter 200 mm with velocity 2 m/s. A heater in the pipe raises the oil temperature to 68°C which reduces the relative density to 0.867.

(a) Use Appendix 3 to determine the viscosity of the oil upstream and downstream of the heater.

(b) Determine the Reynolds number and the flow regime upstream of the heater.

(c) Determine the oil velocity downstream of the heater.

(d) Determine the Reynolds number and flow regime downstream of the heater.

(a) 0.64 Pas, 0.03 Pas (b) 604, laminar (c) 2.23 m/s (d) 12 893, turbulent

1.11 Fluid with viscosity 0.18 Pas and relative density 0.9 flows in a pipe of diameter 100 mm. Determine the maximum flow rate in order that the flow will be laminar. For this flow rate, draw a neat sketch of the velocity profile in the pipe, clearly indicating the shape of the velocity profile curve and the value of the maximum and mean velocity.

31.4 L/s

1.12 Water at a temperature of 5°C flows vertically downward out of a tap in a stream of diameter 5 mm.

(a) Determine the maximum velocity of the water in order for the flow out of the tap to be initially laminar. (Use Appendix 13 for the water viscosity.)

(b) If the water flowing out of the tap had an initial velocity as determined in (a), determine the distance below the tap outlet when turbulent flow commences. Neglect the change in diameter of the stream, and air friction.

(c) What will the flow regime be between the tap outlet and the position where turbulent flow commences?

(a) 0.608 m/s (b) 56.5 mm (c) transition

1.13 Liquid with viscosity 60×10^{-3} Pas and density 875 kg/m^3 flows at a rate of 5 L/s in a pipe of diameter 75 mm. The pipe then tapers to a smaller diameter. Determine:

(a) the Reynolds number and flow regime upstream of the taper;

(b) the minimum velocity and corresponding pipe diameter downstream of the taper in order that flow is turbulent.

(a) 1238, laminar (b) 11.8 m/s, 23.2 mm

1.14 An air duct has a cross-section which is flat at the top and bottom and semicircular at each side. The overall width of the duct is 750 mm and the height is 250 mm. Air with viscosity 16.7×10^{-6} m^2/s flows with velocity 4 m/s through the duct.

(a) Calculate the Reynolds number.

(b) If the duct were circular in shape with the same cross-sectional area, what would the Reynolds number now be?

(a) 93.4×10^3 (b) 113×10^3

Head loss in pipes and fittings

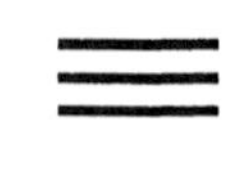

Objectives

On completion of this chapter you should be able to:

- use the Darcy equation to determine the head loss in a pipe;
- determine the friction factor using both Moody diagram and formula;
- determine the head loss through fittings using K factors;
- determine the head loss through a system consisting of a single-diameter pipe and a number of fittings;
- determine the system head equation for a system consisting of a single-diameter pipe and a number of fittings, as well as reservoirs or tanks either vented, or under pressure or vacuum;
- draw the system head equation as a graph;
- determine the fluid power necessary to pump fluid through a system, or the fluid power obtainable from a turbine in the system.

Introduction

In *Thermodynamics and Fluid Mechanics: an Introduction*, the concept of fluid head was introduced. Also the Bernoulli equation relating the various types of fluid head was introduced and used in order to calculate changes to the head of a fluid flowing through a system. When using the Bernoulli equation it was assumed that either the fluid was ideal (no flow loss) or the head loss was given.

In this chapter the Bernoulli equation is revised and then a method for calculating the head loss in pipes and fittings is described. This enables the system head equation to be derived and plotted as a graph.

2.1 FLUID HEAD AND THE BERNOULLI EQUATION

Head is an important concept often used in fluid mechanics because the head of a fluid is directly proportional to its energy. Head is the height to which a fluid will rise as a result of its energy, and has units of metres (m). There are four types of head: pressure head, velocity head, potential head and total head.

Pressure head (h_p) is the height to which a fluid will rise as a result of its pressure. It is given by:

$$h_p = \frac{p}{\rho g} \qquad \left(\frac{\text{Pa}}{\text{kg m}^{-3} \times \text{m s}^{-2}} \right)$$

Velocity head (h_v) is the height to which a fluid will rise as a result of its velocity. It is given by:

$$h_v = \frac{v^2}{2g} \qquad \left(\frac{(\text{m s}^{-1})^2}{\text{m s}^{-2}} \right)$$

Potential head (h) is the height of the centre of mass of a fluid stream above a datum.

Total head (H) is the sum of the pressure, velocity, and potential heads, that is:

$$H = h_p + h_v + h$$

Bernoulli equation

If there are no flow losses, and no fluid machines between positions ① and ②, the toal head of a fluid stream does not change. That is:

$$H = \text{constant, or } H_1 = H_2$$

That is:

$$\frac{p_1}{\rho g} + \frac{v_1^2}{2g} + h_1 = \frac{p_2}{\rho g} + \frac{v_2^2}{2g} + h_2$$

This is the Bernoulli equation for the flow of an ideal fluid.

With real fluids, a flow loss occurs and the flow loss results in a head loss (H_L). The head loss term needs to be included in the right-hand side of the Bernoulli equation, thus:

$$H_1 = H_2 + H_L$$

That is:

$$\frac{p_1}{\rho g} + \frac{v_1^2}{2g} + h_1 = \frac{p_2}{\rho g} + \frac{v_2^2}{2g} + h_2 + H_L$$

This is the Bernoulli equation for the flow of real fluids.

Notes

- This equation is usually applied for flow of liquids. The equation may also be applied (with little error) for flow of gases when there is not more than a 10% change in the density of the gas between ① and ②. Flow under these conditions is known as **incompressible flow.**

- The equation may be applied between any two positions in a fluid flow circuit, provided there is no fluid machinery (such as pumps or turbines) between positions ① and ②. If there is fluid machinery, the equation must be modified further with an additional term which takes into account the head added to, or taken away from, the fluid stream by the machinery. This situation is dealt with later in this chapter under the heading 'System head' (Section 2.7).
- Position ① is the upstream position, whereas position ② is the downstream position, that is, the position further along in *the direction of the fluid flow*. In many cases position ① and/or position ② may be located at the surface of the fluid in a tank or reservoir.
- Pressures p_1 and p_2 may be absolute or gauge, provided that consistency is maintained on both sides of the equation.
- Head loss implies an energy loss, but because of the law of conservation of energy, energy cannot actually be lost or gained. The head loss is in fact a loss of useful energy from the fluid stream. As a result of the head loss the fluid heats up slightly, that is, the temperature and internal energy of the fluid increase. The increase in temperature is usually very small and undetectable[1] and the heat energy transfers outside the fluid to the atmosphere.

It is suggested that Self-test problem 2.1 be now attempted as a revision exercise in the use of the Bernoulli equation.

Self-test problem 2.1

Liquid (*RD* 1.1) flows through the system shown in Figure 2.1. At ①, the pressure is 50 kPa (gauge) and the velocity is 3 m/s.

(a) Determine the velocity and pressure at ② if losses are neglected.

(b) Determine the velocity and pressure at ② if the head loss between ① and ② is 10% of the total head at ① (using position ② as datum).

(c) Convert the head loss to a pressure drop, and show that the difference in pressure calculated in (a) and (b) is equal to the pressure drop.

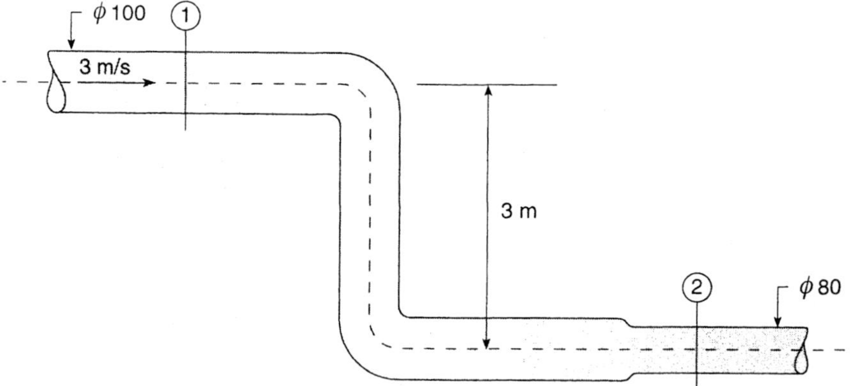

Fig. 2.1

---

[1] Exceptions to this sometimes occur, for example when a spacecraft re-enters the earth's atmosphere. Air friction causes a high increase in temperature at the surface, and special insulation materials are needed in order to prevent heat transfer to the inside of the spacecraft.

2.2 *RELATIONSHIP BETWEEN FLUID POWER AND HEAD*

The relationship between fluid power and head can be expressed by a simple formula:

$$P_f = \dot{m}gH$$

where P_f = fluid power (W)
   $\dot{m}$ = mass flow rate of fluid (kg/s)
   g = gravitational constant (9.81 N/kg or 9.81 m²/s)
   H = head of the fluid (m)

Notes

- In this formula, H is the head of the fluid and usually will be the total head. However, in some cases only one head changes significantly (or is of interest) and then that particular head can be used. For example, in order to find the fluid power associated with a head loss, H_L would be used as the fluid head.
- If the significant head is pressure head, the head would be the pressure head h_p.

  Because $h_p = \dfrac{p}{\rho g}$ and $\dot{m} = \rho \dot{V}$, it follows that $P = \dot{m}gH = \dfrac{\rho \dot{V} g p}{\rho g} = p\dot{V}$.

- Fluid power may increase or decrease the energy of a fluid stream, depending on whether the fluid power is associated with a pump or a turbine. For a pump, the fluid power of the pump increases the energy of the fluid, whereas for a turbine, the fluid power of the turbine decreases the energy of the fluid. A head loss decreases the useful energy of a fluid and therefore reduces the total head.
- Fluid power is *not* the shaft power of a fluid machine (pump or turbine) because no fluid machine is 100% efficient. If P is the shaft power and η is the efficiency, then:

  for a pump, $\eta = \dfrac{P_f}{P}$ and for a turbine, $\eta = \dfrac{P}{P_f}$.

Example 2.1

When a fluid (*RD* 0.92) flows at a rate of 5 L/s through a system, the head loss is 12 m. Determine the loss of fluid power caused by the head loss.

Solution

The mass flow rate $\dot{m}$ = 5 × 0.92 = 4.6 kg/s

$$P_f = \dot{m}gH$$

Substituting,

$$P_f = 4.6 \times 9.81 \times 12 = \textbf{541.5 W}$$

2.3 *HEAD LOSS IN A PIPE*

There are many formulas available for calculating the head loss when a fluid flows through a pipe. One of the most widely used is the Darcy formula (also known as the Darcy–Weisbach formula). This formula is based on the assumption that the head loss is directly proportional to the velocity head. That is:

$$H_L = \text{factor} \times h_v = \text{factor} \times \frac{v^2}{2g}$$

The factor in this formula can be expressed in terms of three variables: the length of pipe, the diameter of the pipe, and a term to take into account the roughness of the pipe and the flow regime. This term is known as the friction factor. Hence the Darcy formula can be written:

$$H_L = f \frac{L}{d} \frac{v^2}{2g}$$

Head loss in a pipe (2.1)

where
$$H_L = \text{head loss of the fluid (m)}$$
$$f = \text{friction factor (dimensionless)}$$
$$L = \text{length of the pipe (m)}$$
$$d = \text{diameter of the pipe (m)}$$
$$v = \text{velocity of the fluid (m/s)}$$
$$g = \text{gravitational constant (9.81 m/s}^2 \text{ or 9.81 N/kg)}$$

Example 2.2

Water flows at a rate of 20 L/s through a pipe of diameter 100 mm and length 1 km. The friction factor is 0.02. Determine:
(a) the head loss;
(b) the pressure drop resulting from the head loss.

Solution

(a) $v = \dfrac{\dot{V}}{A} = \dfrac{0.02}{A(0.1)} = 2.55 \text{ m/s}$

The head loss can be calculated from Formula 2.1:

$$H_L = f \frac{L}{d} \frac{v^2}{2g}$$
$$= 0.02 \times \frac{1000}{0.1} \times \frac{2.55^2}{19.62}$$
$$= \mathbf{66.1 \ m}$$

(b) The relationship between head loss and pressure drop can be determined from the formula:

$$p = \rho g h$$

Because the drop in pressure (Δp) results from a head loss, this formula can be written:

$$\Delta p = \rho g H_L$$
$$\therefore \Delta p = 10^3 \times 9.81 \times 66.1$$
$$= 648 \times 10^3 \text{ Pa}$$
$$= \mathbf{648 \ kPa}$$

Self-test problem 2.2

Petrol (*RD* 0.75) flows through rubber hose of length 12 m and internal diameter 25 mm. The flow rate is 1.5 L/s and the friction factor is 0.018. Determine:

(a) the head loss;
(b) the pressure drop as a result of the head loss;
(c) the fluid power loss.

Example 2.3

Draw a graph of head loss against velocity for the system given in Example 2.2 for water velocity varying between 0 and 5 m/s. Assume a constant friction factor of 0.02.

Solution

From Example 2.2:

$$H_L = f \frac{L}{d} \frac{v^2}{2g}$$

$$= 0.02 \times \frac{1000}{0.1} \times \frac{2.55^2}{19.62}$$

$$= 10.19 v^2$$

The head loss may now be tabulated for various velocities by substituting in the above equation:

v (m/s)	0	1	2	3	4	5
H_L (m)	0	10.2	40.8	91.7	163	255

These points may be plotted as shown in Figure 2.2 on page 24.

Effect of velocity on the head loss in a pipe

From the above example it can be seen that if the friction factor is constant, the head loss varies as the square of the velocity. That is, the head loss–velocity graph is a parabola. Since the fluid power loss varies directly with the head loss, it is clear that high fluid velocities should be avoided in order to keep the head loss and fluid power loss at reasonable levels. For example, a change in fluid velocity of only 1 m/s from 4 m/s to 5 m/s increases the head loss from 163 m to 255 m, which is an increase of 56%.

2.4   *ESTIMATING THE FRICTION FACTOR*

The use of the Darcy formula to calculate the head loss in a pipe is straightforward if the friction factor is given (as in Example 2.2 and Self-test problem 2.2). However, calculating the friction factor for a fluid system involves a more complex procedure because the friction factor depends on a number of variables. These can be grouped under three headings:

- nature of the fluid (in particular its viscosity and density);
- nature of the pipe (in particular its surface roughness and diameter);
- flow regime (whether laminar, turbulent or transition) and the fluid velocity.

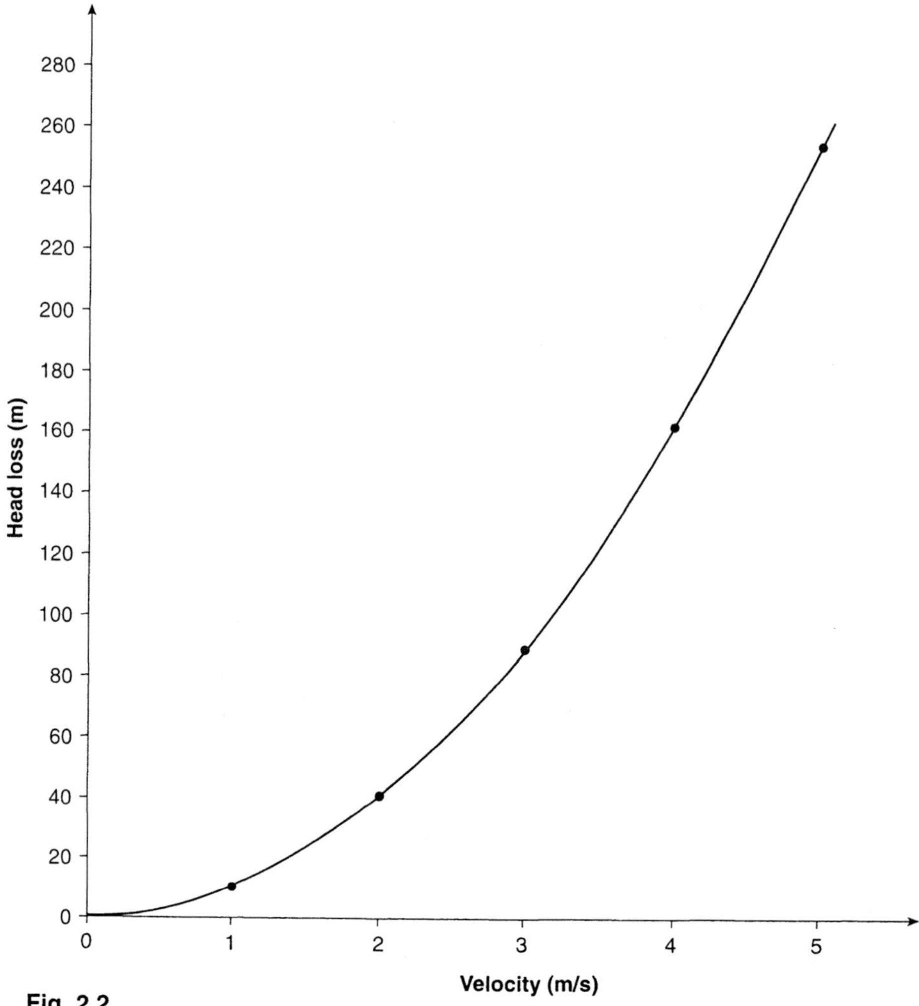

Fig. 2.2

Many of the variables that affect the friction factor appear in the Reynolds number so the Reynolds number is a key variable for determining the friction factor. The first step is therefore to calculate the Reynolds number and hence ascertain the flow regime. Depending on the flow regime and the value of the Reynolds number, the friction factor can be estimated using either a graph or a formula. The friction factor thus determined should be regarded as the best estimate as the friction factor cannot be determined precisely.

The method of determining the friction factor according to the flow regime is now described.

Laminar flow

When flow is laminar, the friction factor is almost independent of the pipe material and the surface roughness. This is because in laminar flow, most of the shearing action (and resulting friction) between layers of fluid takes place away from the wall. Hence the nature of the wall surface has little effect on the friction factor. For laminar flow the friction factor varies inversely with the Reynolds number and the formula is:

$$f = \frac{64}{Re}$$

Friction factor—laminar flow (2.2)

As Reynolds number is dimensionless, the friction factor as calculated by this formula is dimensionless (as it should be).

Self-test problem 2.3

Kerosene at a temperature of 25°C flows through drawn tube of diameter 10 mm at a rate of 20 L/h. Determine the head loss per metre length of pipe. The relative density of kerosene may be taken as 0.78 and the viscosity may be obtained from Appendix 3.

Turbulent flow

When turbulence occurs, the pipe roughness needs to be taken into account when determining the friction factor. This is because there is a very steep velocity gradient at the wall of the pipe, so much of the shearing action (and resulting friction) takes place within the confines of the surface roughness protuberances. The important factor is not the size of the protuberances (known also as the absolute roughness) but the relative roughness defined as:

$$\varepsilon_R = \frac{\varepsilon}{d}$$

Relative roughness (2.3)

where　ε_R　= relative roughness (no units—dimensionless)
　　　　ε　= height of the surface roughness protuberances or absolute roughness (mm)
　　　　d　= pipe diameter (mm)

Note　Relative roughness is dimensionless because the pipe diameter and surface roughness are expressed in the same units (either mm or m). A common error is to express the surface roughness in mm and the pipe diameter in m which leads to an *incorrect* result for relative roughness.

The height of the roughness protuberances is measured as shown in Figure 2.3.

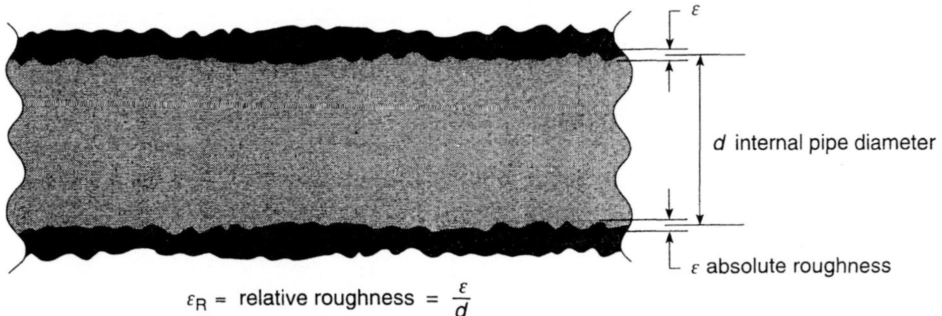

Fig. 2.3　*Absolute and relative roughness*

The absolute roughness for various pipe materials is given in Appendix 4.

Notes

- While it is impossible to produce a perfectly smooth surface, the surface of extruded tubes is so smooth that for the purpose of head loss calculations these tubes can be considered to have a surface roughness of zero.
- Absolute roughness values quoted are for new, clean pipes. It can be expected that after a period of time, corrosion, erosion or scaling may occur. Therefore when designing a fluid system, it is good practice to increase the roughness values to allow for this.

Friction factor diagram

A diagram may be used to determine the friction factor. This diagram is known as the Moody diagram and is given in Appendix 5. In the diagram, friction factor is plotted on the y-axis against Reynolds number on the x-axis (both on a logarithmic scale). A family of curves is shown, each for a different value of the relative roughness.

Notes

- The diagram can be used for laminar as well as turbulent flow. Laminar flow occurs in the region in which $Re < 2000$ (on the left-hand side of the diagram). However, because of the inherent difficulty in reading a diagram accurately, use of the formula leads to a more precise result.
- The diagram cannot be used in the critical region ($2000 < Re < 4000$) as shown by the cross-hatched area. In this region the friction factor does not have a value which can be correlated with the Reynolds number.
- Pipes that have a very low relative roughness can be considered to be smooth and are shown by the lower curved line on the diagram, which is marked accordingly.
- A table of absolute roughness values appears inset in the diagram. It is the same as the table in Appendix 4.
- A broken curved line is shown on the diagram, and to the right of this line the curves flatten out and become horizontal (or nearly so). This region can be termed the constant zone, because the friction factor does not vary significantly with Reynolds number (for a given pipe). When this is the case, the head loss curve will be an exact parabola.
- Between the broken curved line and the curve for a smooth pipe is a region that can be termed the variable zone because the friction factor varies significantly with Reynolds number (for a given pipe). This region is also known as the transition zone.

Friction factor formula

A formula for calculating the friction factor in turbulent flow is the Moody formula:

$$f = 0.0055 \left[1 + \left(20\,000\varepsilon_R + \frac{10^6}{Re} \right)^{\frac{1}{3}} \right]$$

Friction factor— turbulent flow (2.4)

Notes
- This formula is not valid for laminar flow.
- It does not give an exact answer and the Moody diagram is more accurate. However, the maximum difference is usually less than 3% (refer to Self-test problem 2.4). Because of the inherent difficulty in reading a diagram accurately, and because the formula error is usually small, in practice the formula may well give a more precise result.
- In this book, in order to obtain consistent answers all head loss calculations are based on a friction factor as calculated by the formula.
- A great advantage of the formula compared to the diagram is that it can be programmed into a computer or programmable calculator, which cannot be done with the diagram.
- If ε_R is written as a number $\times$ 10^{-3} and Re is written as a number $\times$ 10^6, substitution in the formula is simplified (refer to Example 2.4 part (b)).

Example 2.4

Liquid with relative density 0.9 and viscosity 0.06 Pas flows in a cast-iron pipe of length 100 m and diameter 120 mm. Determine the friction factor using both diagram and formula methods, and the head loss using the formula friction factor value when the velocity is:

(a) 1 m/s
(b) 3 m/s.

Solution

(a) v = 1 m/s

$$Re = \frac{vd\rho}{\mu}$$

$$= \frac{1 \times 0.12 \times 900}{0.06}$$

$$= 1800$$

$\therefore$ flow is laminar

Using the Moody diagram as shown in Figure 2.4 (page 29), f = **0.035**

Using Formula 2.2, $f = \dfrac{64}{Re} = \dfrac{64}{1800} = $ **0.0356**

There is a small difference due to the difficulty in reading the diagram exactly. Using the value obtained from the formula, the head loss can be calculated using the Darcy formula:

$$H_L = f \frac{L}{d} \frac{v^2}{2g}$$

$$= 0.0356 \times \frac{100}{0.12} \times \frac{1^2}{19.62}$$

$$= \textbf{1.51 m}$$

(b) $v = 3$ m/s

$$Re = \frac{vd\rho}{\mu}$$

$$= \frac{3 \times 0.12 \times 900}{0.06}$$

$$= 5400$$

∴ flow is turbulent

From Appendix 4, the absolute roughness of cast iron is 0.25 mm.

The relative roughness from Formula 2.3 is:

$$\varepsilon_R = \frac{\varepsilon}{d} = \frac{0.25}{120} = 2.083 \times 10^{-3}$$

Using the Moody diagram as shown in Figure 2.4 opposite, $f = \mathbf{0.0395}$

Substituting in Formula 2.4, with $Re = 0.0054 \times 10^6$

$$f = 0.0055 \left[1 + \left(20\,000\varepsilon_R + \frac{10^6}{Re}\right)^{\frac{1}{3}}\right]$$

$$= 0.0055 \times \left[1 + \left(20 \times 2.083 + \frac{1}{0.0054}\right)^{\frac{1}{3}}\right]$$

$$= 0.0055 \times \left[1 + (41.7 + 185.2)^{\frac{1}{3}}\right]$$

$$= \mathbf{0.039}$$

Thus there is only a small difference (about 1%) between the calculated value and the value obtained from the diagram. Using the value obtained from the formula, the head loss can be calculated using the Darcy equation:

$$H_L = f\,\frac{L}{d}\,\frac{v^2}{2g}$$

$$= 0.039 \times \frac{100}{0.12} \times \frac{3^2}{19.62}$$

$$= \mathbf{14.9\ m}$$

Self-test problem 2.4

(a) Determine the friction factor using both the Moody diagram and the Moody formula, and the percentage difference, for the following cases:
  - (i) $\varepsilon_R = 0.01$, $Re = 10^7$
  - (ii) $\varepsilon_R = 0$ (smooth pipe), $Re = 10^5$
  - (iii) $\varepsilon_R = 0.002$, $Re = 10^4$

(b) If in Self-test problem 2.3, the flow rate of the kerosene is increased to 0.4 L/s, determine what the head loss per metre length of pipe would now be. Use both the formula and the Moody diagram to determine the friction factor, but base the head loss calculation on the value obtained from the formula.

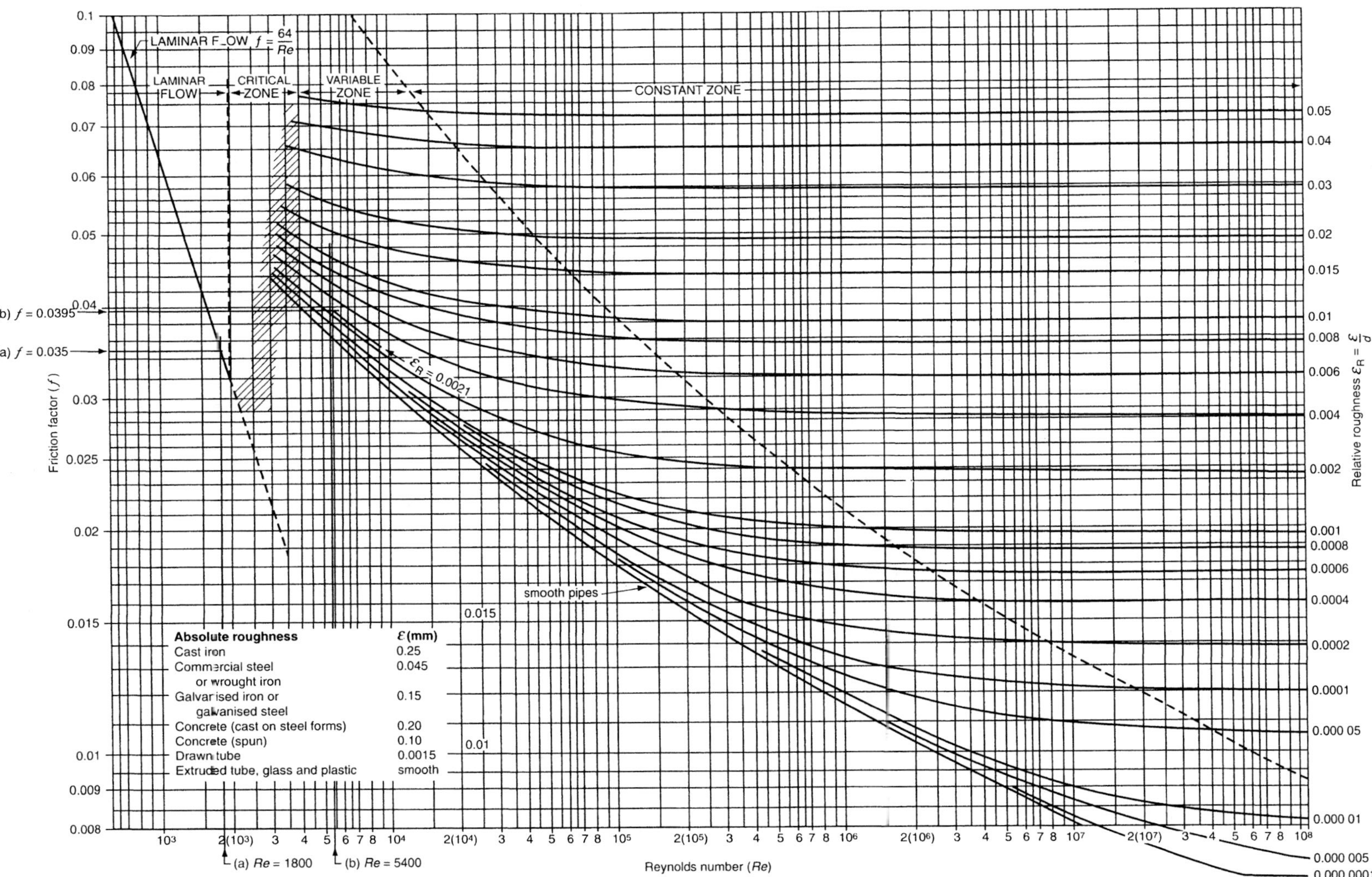

Fig. 2.4

2.5 *HEAD LOSS IN FITTINGS*

Fittings are attachments in a fluid system such as elbows, tees, contractions or enlargements, filters and valves. A fitting in a fluid flow system causes a head loss additional to that due to friction in the pipe or tube. The head loss in fittings is sometimes described as 'minor losses' because in a system with a long length of pipe the head loss in the fittings is usually small compared to the head loss in the pipe. However, if the pipe is relatively short, the head loss in the fittings can be as great or greater than the head loss in the pipe.

The head loss in a fitting is in some part due to fluid friction in the fitting itself, but in the greater part because of turbulence and eddies which occur because the fitting interferes with the smooth flow of fluid. The head loss can be small if the fitting is designed and constructed in accordance with good fluid flow practice, or considerable if the fitting severely interferes with the smooth flow of the fluid. As an example, consider the flow of fluid around an elbow (or bend) as shown in Figure 2.5.

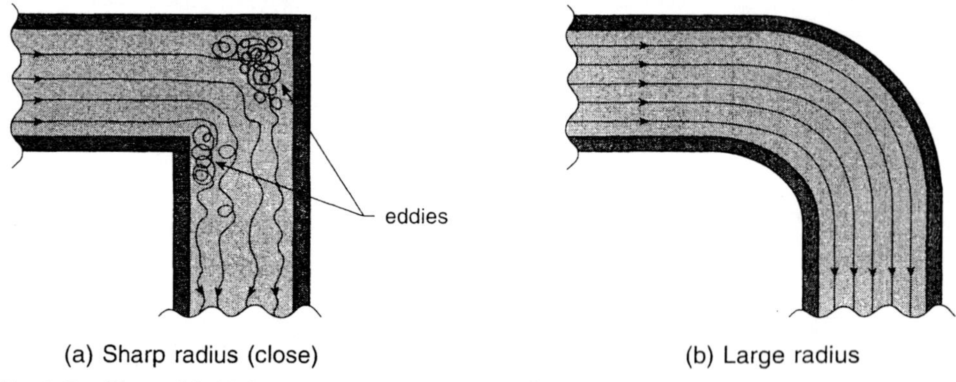

(a) Sharp radius (close)   (b) Large radius

Fig. 2.5   *Flow of fluid through an elbow (or bend)*

For the sharp radius (or close) bend, turbulence and eddies occur at the corners. Like any moving particles with mass and inertia, particles of a fluid cannot follow the exact contour of the sharp corners on both the inside radius and the outside radius of the bend. When the bend has a large radius, these eddies do not occur. However, the larger the radius, the longer is the bend and the greater the head loss due to friction. Since the head loss due to eddies and turbulence is greater than the frictional head loss, a large-radius bend has a lower flow loss than the close bend (for the same diameter and material). The questions that logically follow are: Why not use large radius bends all the time? Why use close bends in a fluid system? There are two reasons why close bends are often used:

- Close bends take up less space and make for a more compact installation than large-radius ones.
- Large-radius fittings are usually more costly than small radius ones. The larger the radius, the more material is required and the more expensive the dies or moulds needed to produce the fitting. Also it is difficult and costly to remove sharp internal corners and provide smooth curves in the internal recesses of a fitting.

Estimating the head loss in fittings

As is the case with a pipe, the head loss in a fitting is directly proportional to the velocity head. That is:

$$H_{\mathrm{L}} = \text{factor} \times h_v = \text{factor} \times \frac{v^2}{2g}$$

The factor is simply known as the *K* factor (*K*). That is:

$$H_{\mathrm{L}} = K \frac{v^2}{2g}$$

Head loss in a fitting (2.5)

where
H_{L} = head loss due to the fitting (m)
K = *K* factor for the fitting (dimensionless)
v = velocity of the fluid (m/s)

When there are a number of fittings in series in a fluid flow circuit the head loss may be written:

$$H_{\mathrm{L}} = \Sigma K \frac{v^2}{2g}$$

Head loss in a number of fittings (2.6)

where ΣK is the sum of the *K* factors for all the fittings. Therefore for a fluid system with a pipe and a number of fittings, the head loss is:

$$H_{\mathrm{L}} = \left(f\frac{L}{d} + \Sigma K \right) \frac{v^2}{2g}$$

Head loss in a fluid system (2.7)

The *K* factor for different types of fittings can be obtained from fluid mechanics data handbooks or from manufacturers' data. It will be found that the *K* factor depends on the size of the fitting and reduces as the size of the fitting increases. Also there can be considerable variation in the value of the *K* factor quoted for a given type of fitting, depending on the manufacturer and the method of manufacture.

A simplified table of typical *K* factors for common fittings is given in Appendix 6 and is used for all problems in this book. The table is based on a nominal size of 100 mm. The values quoted should be regarded as typical values and they should not be used where a more precise estimate of the head loss is required.

Notes

- For a valve, the head loss, and hence the *K* factor, depends on the valve opening. That is to say, as a fully open valve is closed, the *K* factor increases until it reaches infinity in the fully closed position. When the valve is fully open the *K* factor for the valve is at a minimum value which depends on the design and construction of the valve. For example, a fully open gate valve has a much smaller *K* factor than a fully open globe valve because of the much smoother flow path. Similarly a hinged check valve has a lower *K* factor than a ball or poppet type for the same reason.
- The *K* factor table in Appendix 6 includes an estimate of the *K* factor for a partly open globe valve. These values may be useful in the design of a system if the globe valve is used to regulate the flow, and the system is often operating with this valve in a partly open position. No values are given for a partly open gate valve because this valve should be used in the fully on or off position only, and not used for throttling purposes.

- When a pipe exits into a tank, the K factor is 1. That is, all of the velocity head is lost. If the fluid flows from a tank into a pipe and the entrance is sharp, the K factor is 0.5. If the entrance is rounded or tapered the K factor will be less than 0.5. For a well-rounded entrance, the K factor is about one-tenth the value for a sharp entrance, or about 0.05.
- In the case of a foot valve with strainer, the K factor given is based on the assumption that the strainer is clean. As strainers (or filters) clog with use, the K factor increases. Hence when designing a system it may be prudent to increase the K factor of any filters or strainers, to allow for the increased head loss that can occur with use.

Example 2.5

Water at 25°C is pumped at a rate of 20 L/s from a pond to a tank as shown in Figure 2.6. The pipe is 100 mm in diameter, length 60 m, and is made of galvanised steel. In addition to the fittings shown there are 6 screwed sockets. For design purposes the globe valve is taken to be in the 3/4 open position. To allow for clogging, the K factor of the foot valve and strainer is to be increased by one-third. Calculate the head loss.

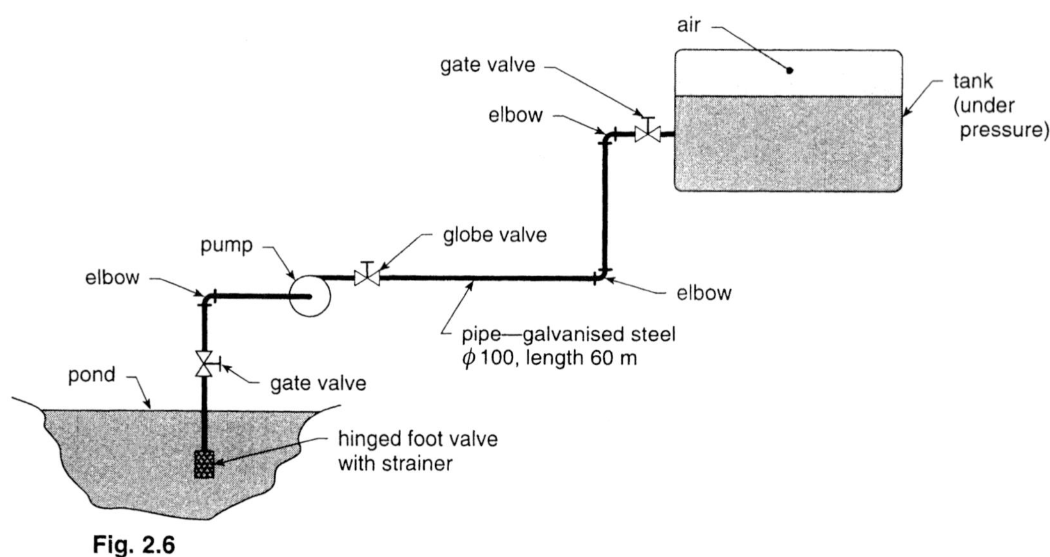

Fig. 2.6

Solution

The velocity is $\upsilon = \dfrac{\dot{V}}{A} = \dfrac{20 \times 10^{-3}}{A(0.1)} = 2.55$ m/s

From Appendix 13, for water at 25°C, $\mu = 0.9 \times 10^{-3}$ Pas and $\rho = 997$ kg/m³. The Reynolds number is:

$$Re = \frac{\upsilon d \rho}{\mu} = \frac{2.55 \times 0.1 \times 997}{0.9 \times 10^{-3}} = 0.282 \times 10^6$$

From Appendix 4, for galvanised steel, $\varepsilon = 0.15$ mm

$$\therefore \varepsilon_R = \frac{0.15}{100} = 1.5 \times 10^{-3}$$

Using Formula 2.4,

$$f = 0.0055 \left[1 + \left(20000\varepsilon_R + \frac{10^6}{Re} \right)^{\frac{1}{3}} \right]$$

$$= 0.0055 \times \left[1 + \left(20 \times 1.5 + \frac{1}{0.282} \right)^{\frac{1}{3}} \right]$$

$$= 0.0232$$

From the Moody diagram, with $Re = 2.82 \times 10^5$ and $\varepsilon_R = 0.0015$:

$$f = 0.0225$$

This is about 3% lower than the calculated value. Using the calculated value:

$$f\frac{L}{d} = 0.0232 \times \frac{60}{0.1} = 13.92$$

The fittings may now be tabulated with their K factors as determined from Appendix 6.

Fitting	K factor	Number off	ΣK
Hinged foot valve with strainer	$3 \times \frac{4}{3} = 4^*$	1	4.0
Gate valve	0.2	2	0.4
Globe valve ($\frac{3}{4}$ open)	8.0	1	8.0
90° elbow	0.6	3	1.8
Screwed socket	0.03	6	0.18
Sudden exit	1.0	1	1.0
		Total	15.38

* The value has been increased by one-third as required.

Thus $f\dfrac{L}{d} + \Sigma K = 13.92 + 15.38 = 29.3$

Using Formula 2.7, $\quad H_L = \left(f\dfrac{L}{d} + \Sigma K \right)\dfrac{v^2}{2g}$

$$= 29.3 \times \frac{2.55^2}{19.62}$$

$$= \mathbf{9.68 \ m}$$

2.6 *EQUIVALENT LENGTH*

The head loss due to a fitting can be calculated using an equivalent length L_E which is defined as the length of pipe that will produce the same head loss as the fitting itself. That is:

$$f \frac{L_E}{d} \frac{v^2}{2g} = K \frac{v^2}{2g}$$

$$\therefore \qquad \boxed{L_E = \frac{Kd}{f}} \qquad\qquad \text{Equivalent length of a fitting (2.8)}$$

If there are a number of fittings in series, then

$$L_E = \frac{\Sigma Kd}{f}$$

Example 2.6

Determine the equivalent length of the fittings used in the system given in Example 2.5 and determine the head loss using this equivalent length.

Solution

In Example 2.5 it was found that $\Sigma K = 15.38$
Also $d = 0.1$ m and $f = 0.0232$

$$L_E = \frac{\Sigma Kd}{f} = \frac{15.38 \times 0.1}{0.0232} = \textbf{66.3 m}$$

That is, the fittings cause the same head loss as 66.3 m of straight pipe. The length of pipe is 60 m, therefore the total equivalent length is: $60 + 66.3 = 126.3$ m.

$$H_L = f \frac{L}{d} \frac{v^2}{2g} = 0.0232 \times \frac{126.3}{0.1} \times \frac{2.55^2}{19.62} = \textbf{9.68 m} \text{ (as before)}$$

2.7 *SYSTEM HEAD*

When a fluid is pumped through a system, the system head is *the total head provided by the pump.* Consider the liquid pumping system shown in Figure 2.7 opposite.

Denoting the surface of the liquid in the pond as position ①, the surface of the liquid in the tank as position ②, and the head provided by the pump as H, the Bernoulli equation may be applied between ① and ②:

$$\frac{p_1}{\rho g} + \frac{v_1^2}{2g} + h_1 + H = \frac{p_2}{\rho g} + \frac{v_2^2}{2g} + h_2 + H_L$$

Rearranging, and making H the subject of the formula:

$$H = \frac{p_2 - p_1}{\rho g} + h_2 - h_1 + \frac{v_2^2 - v_1^2}{2g} + H_L$$

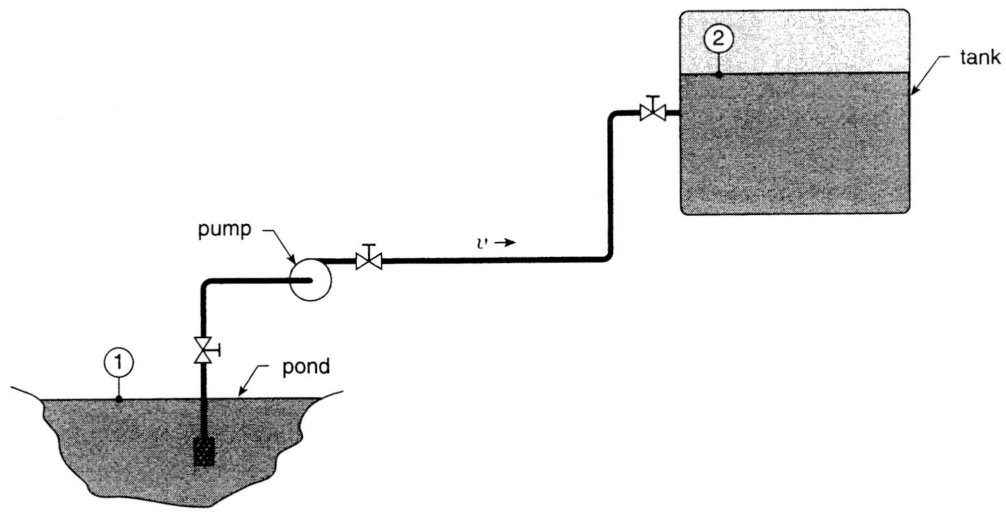

Fig. 2.7 *Liquid pumping system*

Thus the system head has four components:

- Pressure head $= \dfrac{p_2 - p_1}{\rho g}$

  If the pressure at ② is greater than the pressure at ①, the pump must provide a head to overcome this pressure difference.

  If the pressure at ② is equal to the pressure at ①, the pressure head provided by the pump is zero.

  If the pressure at ② is less than the pressure at ①, the pressure head is negative, that is, less head needs to be provided by the pump in order to cause the fluid to flow through the system.

- Potential head $= h_2 - h_1$

  Like pressure head, potential head can have positive, zero or negative value, depending on whether the level at ② is greater than, equal to, or less than the level at ①.

- Velocity head $= \dfrac{v_2{}^2 - v_1{}^2}{2g}$

  If the velocity at ② is greater than the velocity at ①, the pump needs to supply the difference in velocity head. In most cases the velocity head change is small (or zero). For example, in the system shown in Figure 2.6, the tank and the pond have a large surface area so the velocity at these surfaces is negligible and the velocity head change can be taken to be zero. There are cases when the velocity head is large, for example when fluid is pumped through a nozzle, as in a fire hose, or in high velocity jet-cleaning or cutting systems. Then the velocity at exit from the system is much higher than the velocity at inlet, and the velocity head is large and needs to be included in the system head.

- Head loss H_L

  In all real systems where fluid flows there is a head loss. The head loss is always positive and increases the system head.

The four head terms that comprise the system head can be regarded as two subgroups:

- Static head H_{stat}: *Static head is the sum of the pressure head and the potential head.* It is called the static head because the head does not change with flow rate (or velocity).
- Dynamic head H_{dyn}: *Dynamic head is the sum of the velocity head and the head loss.* It is called dynamic head because the head does change with flow rate (or velocity).

Then the system head $H = H_{\text{stat}} + H_{\text{dyn}}$

Example 2.7

For the system given in Example 2.5, the surface of the tank is 5 m above the surface of the pond. The vapour pressure in the tank is 50 kPa (gauge). Determine the system head for the same flow rate, 20 L/s.

Solution

$p_1 = 0$ (using gauge pressures and with the surface at atmospheric pressure)
$v_1 = 0$ (pond surface)
$h_1 = 0$ (datum)
$p_2 = 50$ kPa (gauge)
$v_2 = 0$ (tank surface)
$h_2 = 5$ m
$H_L = 9.68$ m (calculated in Example 2.5)

Applying the system head equation:

$$H = \frac{p_2 - p_1}{\rho g} + h_2 - h_1 + \frac{v_2^2 - v_1^2}{2g} + H_L$$

$$= \frac{50 \times 10^3 - 0}{997 \times 9.81} + 5 - 0 + 0 + 9.68$$

$$= \mathbf{19.79 \ m}$$

2.8   *FLUID POWER AND PUMP POWER*

The system head can be converted to fluid power using the basic fluid power equation:

$$P_f = \dot{m}gH$$

The shaft power can then be determined, knowing the efficiency of the fluid machine (pump or turbine). The method is illustrated in Example 2.8:

Example 2.8

For the system given in Example 2.7 the pump has an efficiency of 65%. Determine the fluid power provided by the pump, and the shaft power necessary to drive the pump.

Solution

Now $\dot{V} = 20$ L/s and from Example 2.8, $H = 19.79$ m. The fluid power is:

$$P_f = \dot{m}gH = 20 \times 0.997 \times 9.81 \times 19.79 \ \text{W}$$

$$= \mathbf{3.87 \ kW}$$

Now for a pump, $\eta = \dfrac{P_f}{P}$

Therefore the shaft power is $P = \dfrac{P_f}{\eta} = \dfrac{3.87}{0.65} = $ **5.96 kW**

2.9 SYSTEM HEAD EQUATION

If the flow rate (or velocity) through a system is known, the system head can be determined. However, in many cases the flow rate is not known. Then it is necessary to obtain a relationship between system head and flow rate (or velocity) as an equation. This equation is known as the **system head equation.**

The relationship can also be shown graphically as a curve in which system head is plotted on the y-axis against flow rate (or velocity) on the x-axis. This graph is known as the **system head curve.** As has been shown, the system head has two components:

- A static component H_{stat} consisting of the pressure head plus the potential head. This component does not change with flow rate (or velocity).
- A dynamic component H_{dyn} consisting of the head loss and a possible velocity head term. If the friction factor is constant, this component varies as the square of the flow rate (and velocity).

Assuming a constant (or nearly constant) friction factor, the system head equation can be written in the following form:

$$H = A + Bv^2$$

System head equation—in terms of velocity (2.8)

This equation can also be written in terms of flow rate as:

$$H = A + C\dot{V}^2$$

System head equation—in terms of flow rate (2.9)

In these equations:

A is a constant and equal to the static head H_{stat} (m).

B and C are factors that depend on the friction factor, length of pipe, diameter of pipe, sum of the K factors and area ratios (if there is a velocity head term in the dynamic head).

The system head curve has the shape shown in Figure 2.8. Two curves are shown, one with velocity plotted on the x-axis, and the other with flow rate plotted on the x-axis.

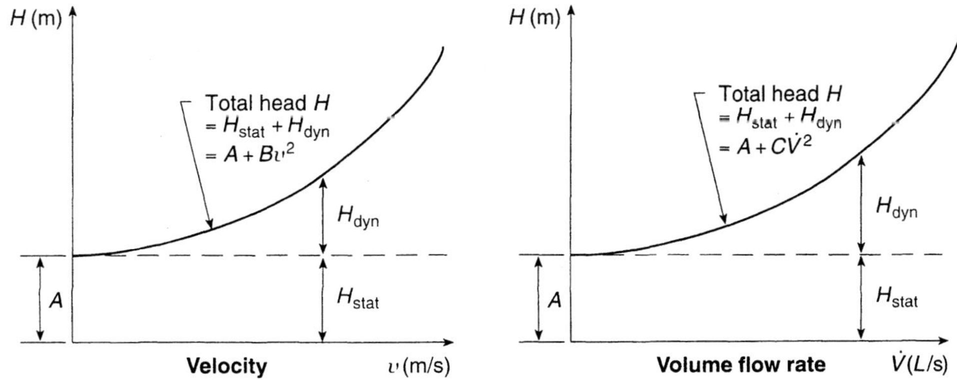

Fig. 2.8 *System head curves*

In the system head curves, constant A is the intercept on the y-axis, and represents the system head when the velocity and flow rate are zero. Constant A will usually have a positive value, and when this is so, the curve starts from a position above the origin on the y-axis. However, constant A can be zero or negative, in which case the curve will start from the origin or below the origin on the y-axis.

Factors B and C determine the steepness or slope of the curve, and the larger their values, the more steeply will the curve rise. If the friction factor is taken to be constant over the range of plotted values (lies in the constant zone in the Moody diagram) factors B and C will be constants and the system head curve will be exactly parabolic.

If the friction factor is not taken to be constant, the values of B and C will not be exactly constant but vary as the velocity and flow rate vary. In this case the system head curve will not be an exact parabola. However, even when the friction factor varies, the variation is usually small and it is sufficiently accurate to use a value of the friction factor based on the mid-range of velocities or flow rates plotted. This is demonstrated in Example 2.9.

Example 2.9

For the system given in Example 2.5 and redrawn in Figure 2.9, the difference in water level between the tank and pond is 3 m. The tank pressure is 30 kPa (gauge).
(a) Show that in the range of flow rates 10–50 L/s, little error results from using a friction factor based on a flow rate of 30 L/s.
(b) Determine the system head equation in the form: $H = A + Bv^2$.
(c) Draw the system head curve for flow rates in the range 0–50 L/s.
(d) Write the system head equation in the form: $H = A + C\dot{V}^2$. Check the accuracy of this equation by using it to calculate the system head at a flow rate of 50 L/s and compare the answer with that derived in (b).

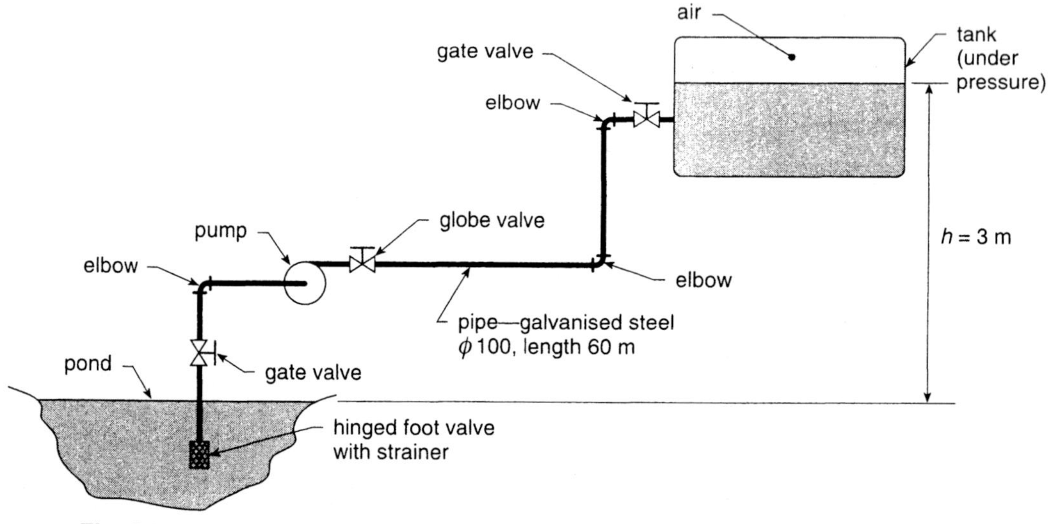

Fig. 2.9

Solution

(a) From Example 2.5:
$$\varepsilon_R = 0.0015, \; L = 60 \text{ m}, \; d = 100 \text{ mm} = 0.1 \text{ m}, \; \mu = 0.9 \times 10^{-3} \text{ Pas},$$
$$\rho = 997 \text{ kg/m}^3, \; \Sigma K = 15.38$$

$$Re = \frac{vd\rho}{\mu} = \frac{v \times 0.1 \times 997}{0.9 \times 10^{-3}} = 110.8 \times 10^3 v$$

$$f = 0.0055 \left[1 + \left(20000\varepsilon_R + \frac{10^6}{Re}\right)^{\frac{1}{3}}\right]$$

$$= 0.0055 \left[1 + \left(30 + \frac{10^6}{Re}\right)^{\frac{1}{3}}\right]$$

Calculating the velocity at each flow rate, and substituting in this formula, the following table can be derived:

$\dot{V}$ (L/s)	10	20	30	40	50
v (m/s)	1.273	2.547	3.820	5.093	6.367
Re	141×10^3	282×10^3	423×10^3	564×10^3	705×10^3
f	0.0238	0.0232	0.0230	0.0229	0.0229

Thus there is little variation in the friction factor, and the use of a mean friction factor $f = 0.023$ will cause little error.

(b) The dynamic head is: $H_{dyn} = \dfrac{v_2^2 - v_1^2}{2g} + \left(f\dfrac{L}{d} + \Sigma K\right)\dfrac{v^2}{2g}$

Now $v_1 = 0$, $v_2 = 0$, $f = 0.023$
Substituting:

$$H_{dyn} = 0 + \left(\frac{0.023 \times 60}{0.1} + 15.38\right)\frac{v^2}{19.62}$$
$$= 1.487 v^2$$

That is, factor $B = 1.487$

The static head is:

$$H_{stat} = \frac{p_2 - p_1}{\rho g} + h_2 - h_1$$
$$= \frac{30 \times 10^3 - 0}{997 \times 9.81} + 3 - 0$$
$$= 6.067 \text{ m}$$

That is, factor $A = 6.067$

Therefore, the system head equation (in terms of velocity) is:

$$\mathbf{H = 6.067 + 1.487 v^2}$$

(c) The system head curve may now be drawn by calculating the velocity at various flow rates, then calculating the system head at each flow rate by substituting in the system head equation. The values thus obtained are given in the table:

$\dot{V}$ (L/s)	0	10	20	30	40	50
v (m/s)	0	1.273	2.55	3.82	5.09	6.37
H (m)	6.07	8.48	15.7	27.8	44.6	66.3

These points can now be plotted as shown in Figure 2.10.

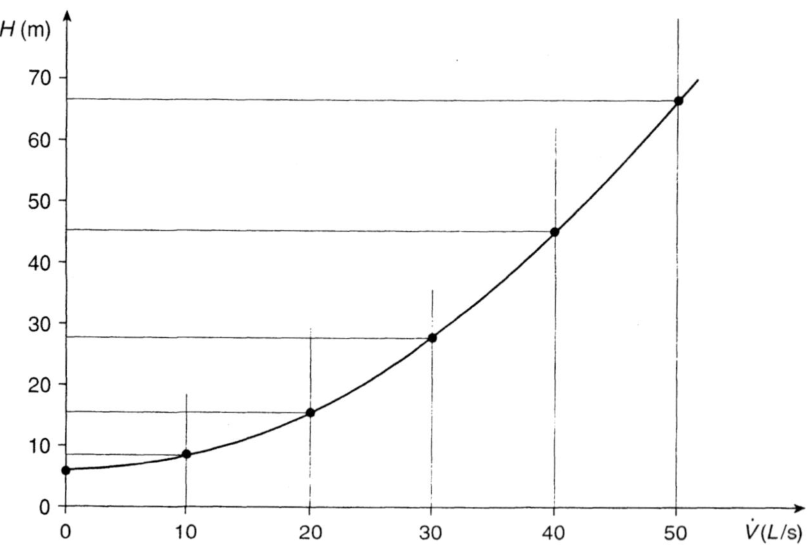

Fig. 2.10

(d) Now $\quad H = 6.067 + 1.487v^2$

and $\quad \dot{V} = vA$

$$\therefore v = \frac{\dot{V}}{A} \quad \left(\frac{\text{m}^3/\text{s}}{\text{m}^2} = \text{m/s}\right)$$

If the flow rate $\dot{V}$ is expressed in L/s, it is necessary to multiply by 10^{-3}

That is, $v = \dfrac{\dot{V}}{A} \left(\dfrac{\text{L/s} \times 10^{-3}}{\text{m}^2}\right) = \dfrac{10^{-3}\dot{V}}{A}$

In this case the pipe has a diameter of 100 mm = 0.1 m

$$\therefore v = \frac{10^{-3}\dot{V}}{A(0.1)} = 0.1273\,\dot{V}$$

Substituting in the system head equation:

$$H = 6.067 + 1.487 \times (0.1273\,\dot{V})^2$$

$$\therefore \boldsymbol{H = 6.067 + 0.0241\,\dot{V}^2}$$

This is the system head equation in terms of flow rate (expressed in L/s).

Check when $\dot{V} = 50$ L/s:

$H = 6.067 + 0.0241 \times 50^2 = \mathbf{66.3\ m}$ (which is the same value as obtained previously).

Self-test problem 2.5

In the system shown in Figure 2.11, liquid with relative density 0.95 and viscosity 5×10^{-3} Pas is pumped from the lower vented tank to the upper tank which is at a pressure of 20 kPa (gauge). The pipe is made of drawn copper and is 30 mm in diameter and 6 m long.

All valves are open.

(a) Determine the system head equation in the form: $H = A + Bv^2$. Assume a constant friction factor based on a velocity v of 2 m/s.

(b) Draw the system head curve for velocities v in the range 0–3 m/s in steps of 0.5 m/s.

(c) Write the system head equation in the form: $H = A + C\dot{V}^2$ where $\dot{V}$ is in L/s. Check the accuracy of this equation by using it to calculate the system head at a flow rate of 2 L/s, and compare the answer with that using the system head curve derived in (b).

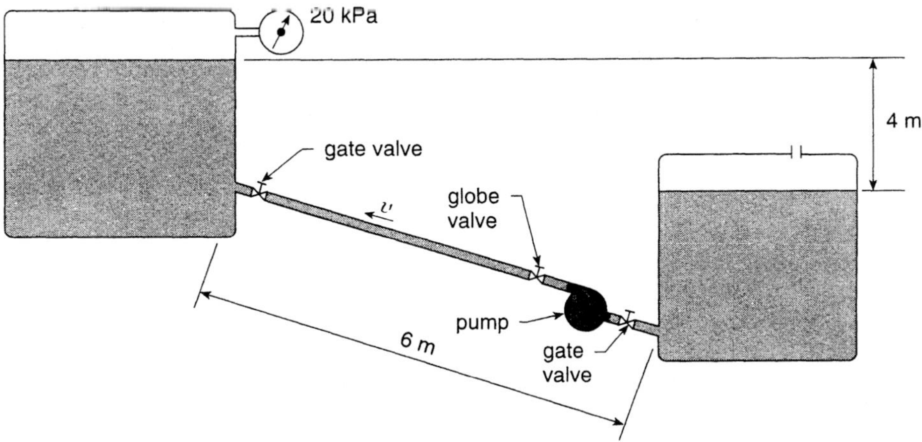

Fig. 2.11

Summary

When an ideal fluid flows through a system so that there is relatively small change in density (incompressible flow) the Bernoulli equation may be used to relate the changes in pressure, velocity and elevation. The equation can be modified for real fluids by including a head loss term on the right-hand side.

Head loss has two components, a head loss due to pipe friction and a head loss due to induced eddies. The head loss that results from pipe friction can be calculated using the Darcy formula:

$$H_L = f \frac{L}{d} \frac{v^2}{2g}$$

To calculate the friction factor, the flow regime must be determined. If flow is laminar, $f = \dfrac{64}{Re}$ and the friction factor is essentially independent of the roughness of the pipe and depends on the Reynolds number only. If the flow is turbulent, the friction factor depends on the Reynolds number as well the surface roughness. The surface roughness is expressed as a relative roughness, that is the absolute roughness (or height of the surface protuberances) divided by the pipe diameter. The friction factor can then be calculated using the Moody formula:

$$f = 0.0055 \left[1 + \left(20000\varepsilon_R + \frac{10^6}{Re} \right)^{\frac{1}{3}} \right]$$

The friction factor can also be obtained graphically using the Moody diagram.

There is also a head loss resulting from induced eddies and friction whenever fluid flows through fittings, or through a sudden enlargement or contraction. This head loss can be estimated by use of K factors in the formula:

$$H_L = \Sigma K \frac{v^2}{2g}$$

The K factor for a fitting depends on the type and size of fitting, the surface roughness, corner sharpness and the extent to which the fitting induces turbulence and sudden changes in the fluid flow direction. The more rapid the change in direction, the higher is the K factor. The larger the fitting, the smaller is the K factor. A simplified table of K factors is given in Appendix 6.

The head loss through a system consisting of a single-diameter pipe and a number of fittings can be calculated using the combined formula:

$$H_L = \left(f \frac{L}{d} + \Sigma K \right) \frac{v^2}{2g}$$

When a fluid flows through a system and the flow rate is given or known, the Bernoulli equation with a head loss term can be used to determine the changes that occur to the fluid between any two points. However, if the flow rate is not known, the Bernoulli equation cannot be used directly and an iterative (trial-and-error) approach is needed. The method of doing this is outlined in Chapter 3.

When a fluid is pumped through a system, the total head supplied by the pump is called the system head. The system head is the sum of the static and dynamic heads, the difference between them being that the static head does not change with flow rate (or velocity), whereas the dynamic head does. The static head is the sum of the pressure and potential head change between the initial and final condition of the fluid, and is given by:

$$H_{stat} = \frac{p_2 - p_1}{\rho g} + h_2 - h_1$$

The dynamic head is the sum of the velocity head change and the head loss between the initial and final condition of the fluid, and is given by:

$$H_{dyn} = \frac{v_2^2 - v_1^2}{2g} + \left(f \frac{L}{d} + \Sigma K \right) \frac{v^2}{2g}$$

Often the velocity head term is negligible and when this is so, the dynamic head is the head loss.

The fluid power necessary to pump a fluid through a system is given by the general formula expressing the relationship between fluid head and power:

$$P_f = \dot{m}gH$$

In this equation the head H is the total system head.

If the total head is expressed as a pressure head (or if the pressure head is the only significant head), the fluid power can be expressed by an alternative formula:

$$P_f = p\dot{V}$$

The power input to the pump can be obtained by dividing the fluid power by the efficiency of the pump.

The way in which the system head varies with flow rate or velocity can be expressed by the system head equation, which includes both the static and dynamic head terms. The equation may be shown graphically as the system head curve. If the friction factor is constant, the system head curve is a parabola, otherwise the curve resembles a parabola but is not exactly parabolic. In most cases, even though the friction factor varies with flow rate (and Reynolds number) it is sufficiently accurate to use a constant value of the friction factor based on the mean flow rate, that is, the flow rate approximately midway between the smallest and largest flow rates.

Problems

Note   Use the Appendixes for all required values that are not given; in particular, use Appendix 13 for the properties of water, Appendix 3 for viscosity of other liquids, Appendix 4 for pipe roughness and Appendix 6 for K factors. All answers are based on friction factors calculated using the Moody formula.

2.1 **(a)** Describe briefly the three groupings under which the variables that affect the friction factor can be placed.

(b) What effect does the surface roughness on a pipe have on the value of the friction factor?

2.2 **(a)** Draw a neat diagram showing the flow of a fluid through a close elbow and a long-sweep elbow. Use these diagrams to explain which elbow has the highest head loss.

(b) Briefly explain the reasons why close elbows (or bends) are often used in preference to long-sweep ones, even though the head loss is higher.

2.3 Liquid, relative density 0.82, flows at a rate of 8 L/s in a 50 mm diameter horizontal pipe which is 40 m long. The upstream pressure is 200 kPa and the friction factor is 0.03. Determine the head loss and the pressure at the end of the pipe.

   20.3 m, 36.7 kPa

2.4 A galvanised steel pipe of diameter 30 mm and length 20 m carries water at a temperature of 20°C with velocity 3 m/s. Determine:

(a) the friction factor;

(b) the head loss;

(c) the pressure drop due to friction.

   (a) 0.0319 (b) 9.77 m (c) 95.6 kPa

2.5 Repeat Problem 2.4 if the velocity is reduced to 1.5 m/s.
(a) 0.0328 (b) 2.51 m (c) 24.5 kPa

2.6 Oil with viscosity 75×10^{-6} m²/s and relative density 0.9 is pumped at a rate of 10 L/s through a horizontal steel pipe of diameter 100 mm and length 50 m. The pump has an efficiency of 70%. Determine:
(a) the friction factor;
(b) the head loss;
(c) the pressure drop;
(d) the fluid power;
(e) the pump shaft power.
(a) 0.0377 (b) 1.55 m (c) 13.75 kPa (d) 137.5 W (e) 196.4 W

2.7 Repeat the calculations for Problem 2.6 if the flow rate is increased to 30 L/s.
(a) 0.0379 (b) 14.11 m (c) 125 kPa (d) 3.74 kW (e) 5.34 kW

2.8 A 50 mm diameter pipeline has the following fittings connected in series:
- 1 sudden contraction (from tank)
- 1 ball check valve
- 1 globe valve (open)
- 4 × 90° elbows
- 10 screwed sockets
- 1 sudden enlargement (to tank).

(a) Determine the head loss through these fittings when the flow rate is 5 L/s.
(b) Explain briefly what is meant by the equivalent length of a fitting, and determine the equivalent length of 50 mm diameter pipe with friction factor 0.02 for these fittings.
(a) 4.69 m (b) 35.5 m

2.9 The pipeline in Problem 2.8 has a total length of 54 m of drawn tube and the fluid has a viscosity of 0.9×10^{-6} m²/s. Determine:
(a) the friction factor using the Moody diagram;
(b) the friction factor using the Moody formula;
(c) the total head loss (using the calculated value of the friction factor).
(a) 0.0168 (b) 0.0163 (c) 10.5 m

2.10 A heat exchanger is manufactured from 20 mm extruded tube of total length 12 m, 10 close-return bends and two 90° elbows connected in series. Oil of relative density 0.9 and viscosity 0.012 Pas flows through the exchanger at a rate of 1.2 L/s. Determine:
(a) the head loss through the exchanger;
(b) the overall K factor for the exchanger as a complete unit;
(c) the equivalent length of the exchanger as a complete unit (in terms of 20 mm diameter tube).
(a) 23 m (b) 30.9 (c) 17.1 m

2.11 Kerosene with relative density 0.78 and at a temperature of 15°C is pumped at a rate of 26.5 L/s through a 150 mm diameter steel pipe of length 100 m. There are 2 gate valves (open), 3 elbows, 5 screwed sockets, 1 swing check valve, 1 sudden entrance, and 1 sudden exit. Determine:
(a) the friction factor;
(b) the equivalent length of the fittings;
(c) the total head loss;
(d) the pressure drop;
(e) the fluid power due to the losses.
(a) 0.0198 (b) 42.1 m (c) 2.145 m (d) 16.4 kPa (e) 435 W

2.12 As shown in Figure P2.12 , a shower head is supplied from a main line where the water temperature is 50°C, the pressure is 200 kPa (gauge), and the velocity is negligible. The shower head has 24 holes and the diameter of the water stream leaving each hole is 0.8 mm. The K factor for the shower head is 0.5 and the tube and hose can be considered smooth.

(a) Assuming a friction factor of 0.02, determine the velocity of the water in the hose. *Hint*: Write an equation for the velocity of the water leaving the shower in terms of the velocity of the water in the hose and substitute in the Bernoulli equation to obtain an equation with one unknown (the velocity of the water in the hose).

(b) Using the velocity calculated in part (a), calculate the friction factor to check if the assumed value was reasonable.

(a) 3.7 m/s (b) 0.0201

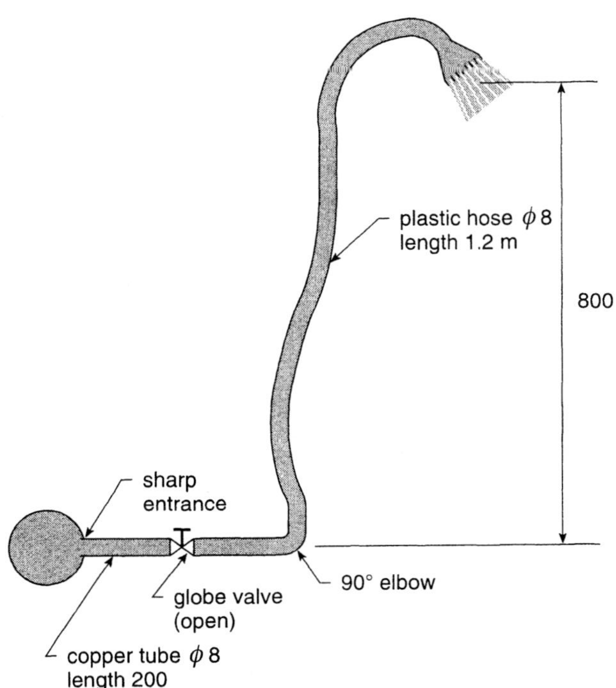

Fig. P2.12

2.13 Water at 25°C is pumped through the system shown in Figure P2.13 from the lower open tank to the upper tank which is at a pressure of 60 kPa. Galvanised steel pipe of diameter 100 mm is used throughout, the length on the suction side of the pump being 2.5 m and on the delivery side 80 m.

The following fittings are used:

Suction side: 2 screwed sockets, 1 hinged foot valve with strainer, 1 elbow
Delivery side: 16 screwed sockets, 1 globe valve (half-open), 2 elbows, 1 gate valve (open) and 1 sudden enlargement (to tank).

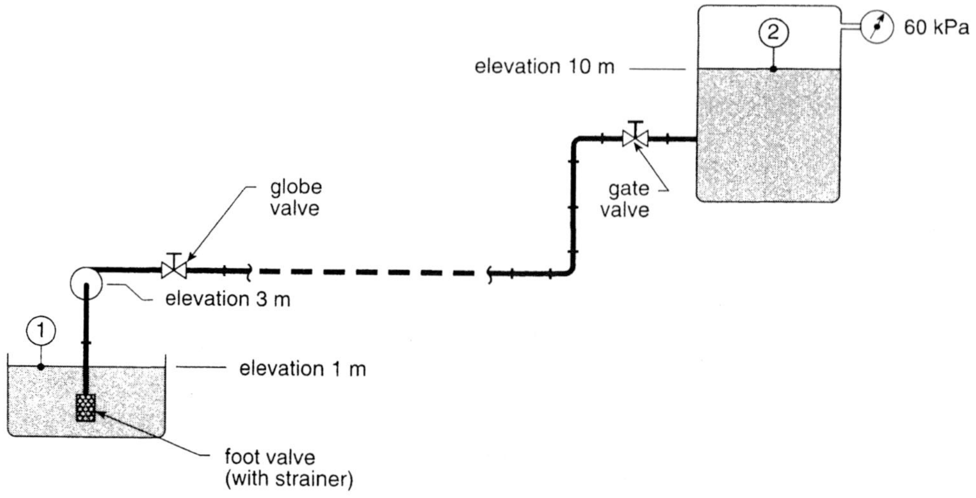

Fig. P2.13

Assume a constant friction factor based on a flow rate of 10 L/s.
(a) Determine the system head equation (in terms of velocity).
(b) Determine the system head equation (in terms of flow rate in L/s).
(c) Determine the system head for flow rates 0–20 L/s in steps of 5 L/s.
(d) Draw the system head curve to scale.
(e) Explain briefly the conditions under which the system head curve is exactly parabolic.

(a) $H = 15.13 + 1.9475v^2$ (b) $H = 15.13 + 0.0316\dot{V}^2$
(c)

$\dot{V}$ (L/s)	0	5	10	15	20
H (m)	15.13	15.9	18.3	22.2	27.8

2.14 Lubricating oil with relative density 0.9 and viscosity 0.05 Pas is pumped at a rate of 20 L/s from a tank at atmospheric pressure to a storage tank at a pressure of 50 kPa (gauge). The increase in elevation is 6 m. The pump has an efficiency of 72% at the given flow rate. The system consists of 60 m of 80 mm diameter steel pipe with the following fittings: 1 sudden entrance, 1 sudden exit, 1 gate valve (open), 1 globe valve (open), 15 screwed sockets and 4 × 90° elbows. Determine:
(a) the static head;
(b) the friction factor (using the Moody formula);
(c) the equivalent length of the fittings;
(d) the dynamic head;
(e) the system head;
(f) the fluid power;
(g) the pump shaft power.

(a) 11.66 m (b) 0.0369 (c) 22.9 m (d) 30.8 m (e) 42.5 m (f) 7.5 kW (g) 10.4 kW

2.15 For the system given in Problem 2.14, assume a constant friction factor of 0.03, and determine:
(a) the system head equation in terms of velocity;
(b) the system head equation in terms of flow rate in L/s;
(c) the system head and fluid power for flow rates of 0 . 5, 10. 15. 20. and 25 L/s.

Also draw a graph of system head and fluid power against flow rate on one sheet of graph paper, using the values calculated in (c).

(a) $H = 11.66 + 1.6845v^2$ (b) $H = 11.66 + 0.0667\dot{V}^2$

(c)

Flow rate (L/s)	0	5	10	15	20	25
System head (m)	11.66	13.3	18.3	26.7	38.3	53.3
Fluid power (kW)	0	0.588	1.62	3.53	6.77	11.8

2.16 In a galvanised steel water supply line 100 m long and 50 mm in diameter, water is pumped from a reservoir to a vented storage tank at an elevation 8 m above it. The following fittings are used: 1 hinged foot valve with strainer, 3 × 90° elbows, 2 gate valves (open), 20 screwed sockets, 1 sudden exit. The water viscosity is 1 × 10^{-6} m²/s and the density is 1000 kg/m³.

(a) Calculate the friction factor at a flow rate of 3 L/s.

(b) Using a constant friction factor as determined in (a), determine the system head equation in terms of velocity.

(c) Using a constant friction factor as calculated in (a), determine the system head equation in terms of flow rate in L/s.

(d) Using a constant friction factor as calculated in (a), determine the system head and fluid power for flow rates of 0, 1.5, 3, 4.5, and 6 L/s.

(e) Draw a graph of system head and fluid power against flow rate on one sheet of graph paper, using the values calculated in (d).

(f) If the efficiency of the pump is 65% when the flow rate is 5 L/s, determine the input power to the pump at this flow rate.

(a) 0.0285 (b) $H = 8 + 3.251v^2$ (c) $H = 8 + 0.8433\dot{V}^2$

(d)

Flow rate (L/s)	0	1.5	3	4.5	6
System head (m)	8	9.90	15.6	25.1	38.4
Fluid power (kW)	0	0.146	0.459	1.11	2.26

(e) 2.195 kW

2.17 Compressed air at an initial temperature of 20°C and a pressure of 150 kPa (gauge) flows at a rate of 2.5 g/s through a system which has 50 m of 12 mm diameter steel pipe and the following fittings: 10 × 90° elbows, 2 gate valves and 1 globe valve (open), 20 screwed sockets, and 3 tees with flow along the line. Incompressible flow may be assumed. Determine:

(a) the velocity of the air in the pipe.
Hint: Assume R for air is 287 J/kgK and use the perfect gas equation to calculate the air density;

(b) the friction factor (use Appendix 3 for the viscosity of air);

(c) the air pressure drop through the system, and show that this is less than 10% of the initial pressure and so the assumption of incompressible flow was valid.

(a) 7.4 m/s (b) 0.0343 (c) 12.8 kPa (<10%)

2.18 In a fire-fighting installation as illustrated in Figure P2.18, the fire hose when full has a diameter of 75 mm. Water leaves the nozzle at the end of the hose at a temperature of 10°C in a stream of diameter 25 mm at an elevation 5 m above the level of the pump. The hose has a length of 50 m and the average height of the roughness protuberances is 0.2 mm. Minor losses are negligible.

(a) Determine the friction factor at a flow rate of 10 L/s.

(b) Determine the system head equation in terms of velocity v.

(c) Using a friction factor based on a flow rate of 10 L/s, complete the following table:

Flow rate (L/s)	0	5	10	15	20
Velocity v (m/s)					
System head (m)					
Fluid power (kW)					
Pressure at pump outlet (kPa gauge)					

(d) Plot these results as four lines on one sheet of graph paper, where flow rate in L/s is on the *x*-axis.

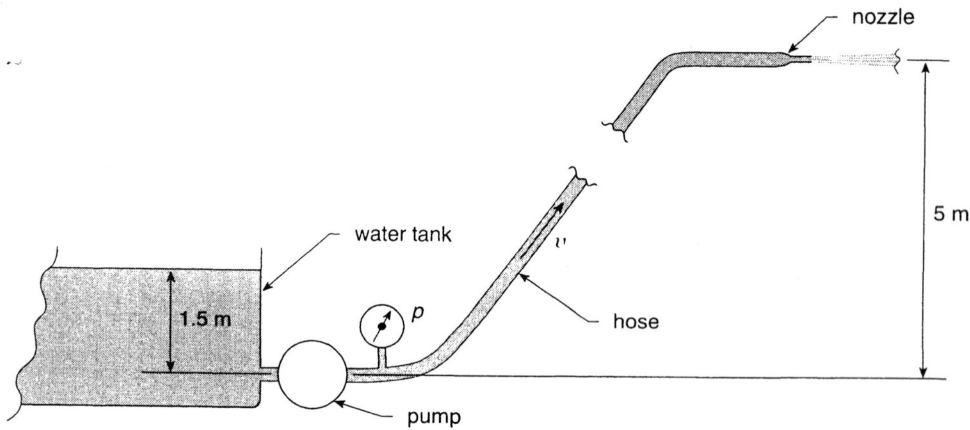

Fig. P2.18

(a) 0.0272 (b) $H = 3.5 + 5.051v^2$

(c)

Flow rate (L/s)	0	5	10	15	20
Velocity v (m/s)	0	1.13	2.26	3.4	4.53
System head (m)	3.5	9.97	29.4	61.7	107
Fluid power (kW)	0	0.489	2.88	9.08	21.0
Pressure at pump outlet (kPa gauge)	49.1	112	300	615	1055

Pipe flow

Objectives

On completion of this chapter you should be able to:

- determine the free-flow rate or head loss from a tank through a single-diameter pipe and a number of fittings, or from the tank to another, either tank being vented or under pressure;
- determine the free-flow rate or head loss from a tank through a number of pipes in series and a number of fittings, or from the tank to another, either tank being vented or under pressure;
- determine the free-flow rate or head loss from a tank through a number of pipes in parallel and a number of fittings, or from the tank to another, either tank being vented or under pressure;
- determine the free-flow rate or head loss from a tank through a number of pipes in a series and parallel combination and a number of fittings, or from the tank to another, either tank being vented or under pressure.

Introduction

In Chapter 2 a method for estimating the head loss in a fluid system consisting of a single-diameter pipe and a number of fittings was given. Assuming incompressible flow (liquids, or small variations in gas density) the Bernoulli equation could be used for real fluid flow situations by including the head loss term. With a known (or given) head loss and pipe diameter, a direct solution to the equation could be obtained, that is, a trial-and-error (or iterative) approach would not be necessary.

In many cases, the head loss, flow rate (or velocity) and pipe diameter may be unknown. In such cases a direct solution is usually not possible and a trial-and-error (or iterative) approach is required. This is because the head loss, flow rate (or velocity) and pipe diameter are all interdependent. For example, in free flow of a fluid from a tank or reservoir through a pipe the head loss depends on the flow rate, and the flow rate depends on the head loss. This interdependence prevents a direct solution when attempting to calculate the flow rate, head loss or pipe diameter needed for a given flow rate.

Not all fluid systems have a single-diameter pipe of the same material. Where pipes of different diameter, or of different materials, are joined in series or parallel, a further complication occurs and the solution is more complex. Often the calculation procedure is simplified by reducing the series or parallel pipes to a single-diameter equivalent pipe. Then the solution follows the same procedure as that needed when the fluid flows through a single-diameter pipe.

3.1 *FLOW THROUGH A SINGLE-DIAMETER PIPE (UNKNOWN FLOW RATE)*

A typical system of this type is illustrated in Figure 3.1.

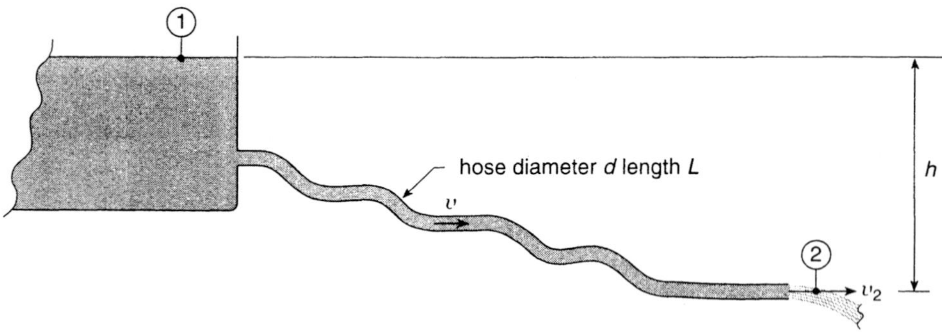

Fig. 3.1   *Flow out of a single-diameter pipe*

Liquid drains through a pipe (or hose) out of a large tank or reservoir at atmospheric pressure.

The Bernoulli equation may be applied between ① and ②, with ② as datum:

$$\frac{p_1}{\rho g} + \frac{v_1^2}{2g} + h_1 = \frac{p_2}{\rho g} + \frac{v_2^2}{2g} + h_2 + H_L$$

Now $p_1 = 0$ (atmospheric), $v_1 = 0$ (large tank), $h_1 = h$
$p_2 = 0$ (atmospheric), $h_2 = 0$ (datum)

Substituting these values in the Bernoulli equation:

$$h = \frac{v_2^2}{2g} + H_L$$

Now $H_L = \left(f\frac{L}{d} + \Sigma K \right) \frac{v^2}{2g}$

where v = velocity through the hose

$$\therefore\, h = \frac{v_2^2}{2g} + \left(f\frac{L}{d} + \Sigma K \right) \frac{v^2}{2g}$$

In practice, the diameter of the stream just beyond the outlet to the hose is somewhat smaller than the diameter of the hose, so v_2 is a little greater than the velocity in the hose v. Usually, this difference is not significant and the assumption may be made that $v_2 = v$. This assumption will be made for all such situations in this chapter.

Hence the equation becomes:

$$h = \frac{v^2}{2g} + \left(f\frac{L}{d} + \Sigma K\right)\frac{v^2}{2g}$$

that is:

$$h = \left(1 + f\frac{L}{d} + \Sigma K\right)\frac{v^2}{2g}$$

Note   This equation should not be memorised but derived using the Bernoulli equation and the Darcy formula according to the system. The equation is different if the tank is under pressure, or when one tank is joined to another tank.

Known friction factor

When the friction factor is known, the solution is direct and is illustrated in Example 3.1.

Example 3.1

Water at 15°C drains from a large tank as shown in Figure. 3.1. The hose has a diameter of 15 mm and a length of 20 m. The outlet to the hose is 5 m below the surface level in the tank. Determine the flow rate through the hose if the friction factor is 0.025.

Solution

As derived above, the equation is:

$$h = \left(1 + f\frac{L}{d} + \Sigma K\right)\frac{v^2}{2g}$$

The only fitting involved is at the exit from the tank (entrance to the pipe) where $K = 0.5$.
Also $h = 5$ m, $L = 20$ m, $d = 15$ mm $= 0.015$ m, and $f = 0.025$

Substituting these values:

$$5 = \left(1 + \frac{0.025 \times 20}{0.015} + 0.5\right)\frac{v^2}{19.62}$$

$$\therefore v = 1.68 \text{ m/s}$$

The flow rate is: $\dot{V} = vA = 1.68 \times A(0.015) \text{ m}^3/\text{s} = \textbf{0.297 L/s}$

Unknown friction factor

When the friction factor is not known, the solution is not direct and requires a trial-and-error approach. In the absence of any other knowledge, a first trial value of $f = 0.02$ is often a good starting point. Usually flow is turbulent so the Moody formula (2.4) may be used to calculate the friction factor. If any doubt exists as to the flow regime, a calculation of the Reynolds number which produces a value greater than 4000 will confirm turbulent flow.

A trial-and-error solution is illustrated in Example 3.2:

Example 3.2

Repeat Example 3.1, but with unknown friction factor. The hose is made of material with absolute roughness 0.01 mm.

Solution

In this case, it is logical to use the values assumed in Example 3.1 as the first trial. In this example it was found that for an assumed value $f = 0.025$, the corresponding velocity was 1.68 m/s. This value of the velocity will now be used as the initial value in Trial 1 to calculate the friction factor.

Trial 1

From Appendix 13, for water at 15°C, $\mu = 1.15 \times 10^{-3}$ Pas, $\rho = 999$ kg/m^3

$$Re = \frac{vd\rho}{\mu} = \frac{1.68 \times 0.015 \times 0.999 \times 10^3}{1.15 \times 10^{-3}} = 0.0219 \times 10^6$$

$$\varepsilon_R = \frac{0.01}{15} = 0.667 \times 10^{-3}$$

Because the Reynolds number is much greater than 4000, flow will be turbulent, so the Moody formula may be used to calculate the friction factor:

$$f = 0.0055 \left[1 + \left(20\,000 \varepsilon_R + \frac{10^6}{Re} \right)^{\frac{1}{3}} \right]$$

Substituting:

$$f = 0.0055 \left[1 + \left(20 \times 0.667 + \frac{1}{0.0219} \right)^{\frac{1}{3}} \right]$$

$$= 0.0055 \left[1 + (13.33 + 45.63)^{\frac{1}{3}} \right]$$

$$= 0.0269 \text{ (and the assumed value } f = 0.025 \text{ in Example 3.1 was reasonable)}$$

As derived previously, the Bernoulli equation reduces to:

$$h = \left(1 + f\frac{L}{d} + \Sigma K \right) \frac{v^2}{2g}$$

With $\Sigma K = 0.5$, $L = 20$ m, $h = 5$ m and $d = 0.015$ m, the equation is:

$$5 = \left(1.5 + \frac{f \times 20}{0.015}\right)\frac{v^2}{19.62}$$

Substituting the value $f = 0.0269$ as calculated:

$$5 = \left(1.5 + \frac{0.0269 \times 20}{0.015}\right)\frac{v^2}{19.62}$$

$$5 = \frac{37.34v^2}{19.62}$$

$$\therefore v = 1.62 \text{ m/s}$$

Therefore, the trial value of 1.68 m/s was a little too high.

Trial 2

Try $v = 1.62$ m/s

$$Re = \frac{vd\rho}{\mu} = \frac{1.62 \times 0.015 \times 0.999 \times 10^3}{1.15 \times 10^{-3}} = 0.0211 \times 10^6$$

ε_R is the same, that is, 0.667×10^{-3}

$$f = 0.0055\left[1 + \left(20\,000\varepsilon_R + \frac{10^6}{Re}\right)^{\frac{1}{3}}\right]$$

$$= 0.0055\left[1 + \left(13.33 + \frac{1}{0.0211}\right)^{\frac{1}{3}}\right]$$

$$= 0.0055\left[1 + (13.33 + 47.32)^{\frac{1}{3}}\right]$$

$$= 0.0271$$

Substituting the value $f = 0.0271$ as calculated:

$$5 = \left(1.5 + \frac{0.0271 \times 20}{0.015}\right)\frac{v^2}{19.62}$$

$$\therefore 5 = \frac{37.63v^2}{19.62}$$

$$\therefore v = 1.615 \text{ m/s}$$

Clearly there is little to be gained from further trials so $v = 1.615$ m/s.

$$\dot{V} = vA = 1.615 \times A(0.015) = 0.285 \times 10^{-3} \text{ m}^3\text{/s} = \textbf{0.285 L/s}$$

 ## *Self-test problem 3.1*

In an industrial process a tank under pressure is connected to a vented tank as shown in Figure 3.2. The liquid has a relative density of 0.95 and a viscosity of 5×10^{-3} Pas. The pipe connecting the two tanks has a diameter of 30 mm, a length of 6 m, and is made of drawn copper. Determine the velocity v of the liquid when all valves are fully open. Two trials only are required, using an initial trial value of $v = 2$ m/s.

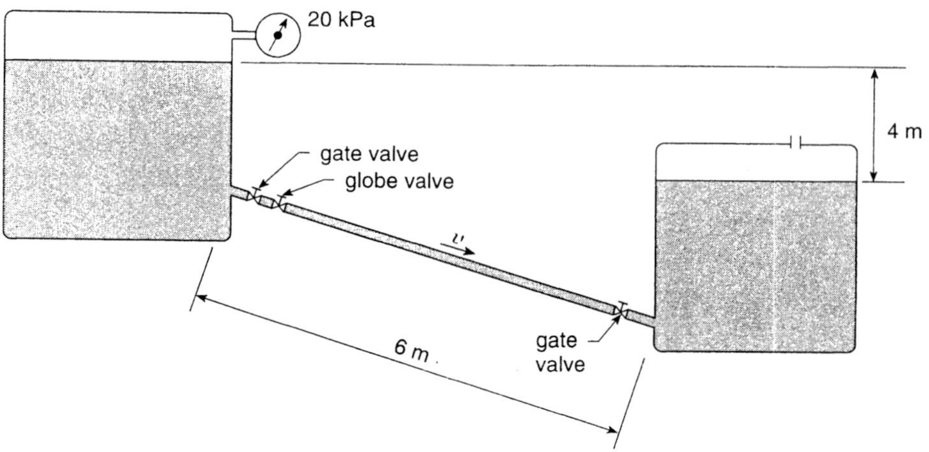

Fig. 3.2

3.2 FLOW THROUGH A SINGLE-DIAMETER PIPE (UNKNOWN PIPE DIAMETER)

A typical system of this type is illustrated in Figure 3.3.

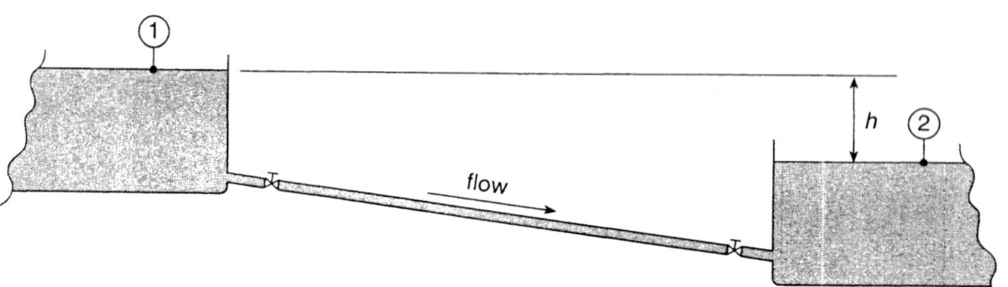

Fig. 3.3 *Flow from one reservoir to another*

Liquid drains through a pipe out of a large tank or reservoir at atmospheric pressure to another tank or reservoir which is below it. The problem is to determine the pipe diameter needed in order to achieve a given flow rate.

The Bernoulli equation may be applied between ① and ②, with ② as datum:

$$\frac{p_1}{\rho g} + \frac{v_1^{\,2}}{2g} + h_1 = \frac{p_2}{\rho g} + \frac{v_2^{\,2}}{2g} + h_2 + H_L$$

Now $p_1 = 0$ (atmospheric), $v_1 = 0$ (large tank), $h_1 = h$
      $p_2 = 0$ (atmospheric), $v_2 = 0$ (large tank), $h_2 = 0$ (datum)

Substituting these values in the Bernoulli equation:

$$h = H_L$$

Now $\qquad H_{\mathrm{L}} = \left(f \dfrac{L}{d} + \Sigma K \right) \dfrac{v^2}{2g}$

where $\qquad v$ = velocity through the pipe

$$\therefore h = \left(f \dfrac{L}{d} + \Sigma K \right) \dfrac{v^2}{2g}$$

Note   This equation should not be memorised, but derived using the Bernoulli equation and the Darcy formula according to the system. The equation is different if one or both of the tanks is under pressure.

Known friction factor

When the friction factor is known, a direct solution is theoretically possible. Because the flow rate is given, the velocity may be expressed in terms of the pipe diameter. This results in one equation with one unknown (the pipe diameter). If minor losses are neglected, the solution is straightforward, but if minor losses are included, the equation may be solved by a trial-and-error approach. The method is illustrated in Example 3.3.

Example 3.3

For the system illustrated in Figure 3.3, the flow rate is to be 10 L/s when the difference in elevation h between the two tanks is 1 m. Both valves are open gate valves. The pipe is 40 m long and the friction factor may be taken as 0.02. Determine the diameter of pipe required (to the nearest mm) if:
(a) minor losses are neglected;
(b) minor losses are included.

Solution

The equation is:

$$h = \left(f \dfrac{L}{d} + \Sigma K \right) \dfrac{v^2}{2g}$$

Now $h = 1$ m, $f = 0.02$, and $L = 40$ m

(a) If minor losses are neglected, $\Sigma K = 0$
    Substituting these values, the equation becomes:

$$1 = \left(\frac{0.02 \times 40}{d} \right) \frac{v^2}{19.62}$$

Also $\qquad \dot{V} = vA$

Now $\qquad \dot{V} = 10 \text{ L/s} = 0.01 \text{ m}^3/\text{s} \text{ and } A = \dfrac{\pi d^2}{4}$

$$\therefore 0.01 = v \times \frac{\pi d^2}{4}$$

$$\therefore v = \frac{0.0127}{d^2} \text{ and } v^2 = \frac{0.1621 \times 10^{-3}}{d^4}$$

Substituting:

$$1 = \left(\frac{0.02 \times 40}{d}\right) \times \frac{0.1621 \times 10^{-3}}{19.62 \times d^4}$$

$$\therefore\ 121\,026 = \frac{0.8}{d^5}$$

$$\therefore\ d^5 = 6.61 \times 10^{-6}$$

$$\therefore\ d = 0.0921 = \textbf{92 mm}\ \text{(to the nearest mm)}$$

(b) If minor losses are included:

$\Sigma K = 0.5$ (sudden entrance) $+$ 0.4 (two open gate valves) $+$ 1.0 (sudden exit) $= 1.9$

Substituting these values, the equation becomes:

$$1 = \left(\frac{0.02 \times 40}{d} + 1.9\right)\frac{v^2}{19.62}$$

Also $\qquad \dot{V} = vA$

As previously obtained, $v^2 = \dfrac{0.1621 \times 10^{-3}}{d^4}$

Substituting:

$$121\,026 = \left(\frac{0.8}{d} + 1.9\right) \times \frac{1}{d^4}$$

This equation may be solved by trial and error.

In (a), when minor losses were neglected, it was found that $d = 92$ mm.

Now when minor losses are included, the diameter will need to be slightly greater. Hence as a first trial, take $d = 95$ mm.

Trial 1

$\qquad d = 95$ mm

The right-hand side of the equation, $RHS = \left(\dfrac{0.8}{0.095} + 1.9\right) \times \dfrac{1}{0.095^4} = 126\,715$

This is greater than the value on the left-hand side: $LHS = 121\,026$. Because the magnitude of the RHS will reduce as d increases, the value of d needs to be increased for the next trial.

Trial 2

$\qquad d = 96$ mm

$$RHS = \left(\frac{0.8}{0.096} + 1.9\right) \times \frac{1}{0.096^4} = 120\,484$$

This value is now lower than the *LHS*, so the precise solution occurs for a value of d lying between 95 and 96 mm. Since Trial 2 produced the closest correlation between the *LHS* and *RHS*, the solution to the nearest mm is d = **96 mm**.

Unknown friction factor

When the friction factor is unknown, a direct solution is not possible. A trial-and-error approach can be adopted where trial values of either the velocity or the diameter may be successively refined. In Example 3.4, trial values of the velocity are refined until the velocity and diameter are determined to the required level of accuracy.

Example 3.4

For the system illustrated in Figure 3.3 (page 54) the liquid is petrol which has viscosity 0.83×10^{-6} m²/s. The flow rate is to be 10 L/s when the difference in elevation h between the two tanks is 1 m. Both valves are open gate valves. The pipe is 40 m long and made of steel. Determine the diameter of pipe required (to the nearest mm). Take as the initial trial value: v = 1 m/s.

Solution
Trial 1

$$v = 1 \text{ m/s}$$

Now
$$\dot{V} = 10 \text{ L/s} = 0.01 \text{ m}^3/\text{s} \text{ and } A = \frac{\pi d^2}{4}$$

$$\therefore 0.01 = 1 \times \frac{\pi d^2}{4}$$
$$\therefore d = 0.113 \text{ m} = 113 \text{ mm}$$

From Appendix 4, ε = 0.045 mm (commercial steel)

$$\varepsilon_R = \frac{0.045}{113} = 0.398 \times 10^{-3}$$

Given v for petrol = 0.83×10^{-6} m²/s

$$Re = \frac{vd}{v} = \frac{1 \times 0.113}{0.83 \times 10^{-6}} = 0.136 \times 10^6 \text{ (and flow is turbulent)}$$

$$f = 0.0055 \left[1 + \left(20\,000\varepsilon_R + \frac{10^6}{Re} \right)^{\frac{1}{3}} \right]$$

$$= 0.0055 \left[1 + \left(20 \times 0.398 + \frac{1}{0.136} \right)^{\frac{1}{3}} \right]$$

$$= 0.0192$$

As derived previously, the Bernoulli equation reduces to:

$$h = \left(f \frac{L}{d} + \Sigma K \right) \frac{v^2}{2g}$$

Substituting $f = 0.0192$, $\Sigma K = 1.9$, $L = 40$ m, $d = 0.113$ m, and $h = 1$ m:

$$1 = \left(\frac{0.0192 \times 40}{0.113} + 1.9\right) \times \frac{v^2}{19.62}$$

$$\therefore v = 1.5 \text{ m/s}$$

Trial 2

Take as the next trial value $v = 1.5$ m/s

$$0.01 = 1.5 \times \frac{\pi d^2}{4}$$

$$\therefore d = 0.092 \text{ m} = 92 \text{ mm}$$

$$\varepsilon_R = \frac{0.045}{92} = 0.49 \times 10^{-3}$$

$$Re = \frac{vd}{v} = \frac{1.5 \times 0.092}{0.83 \times 10^{-6}} = 0.166 \times 10^6$$

$$f = 0.0055 \left[1 + \left(20000\varepsilon_R + \frac{10^6}{Re}\right)^{\frac{1}{3}}\right]$$

$$= 0.0055 \left[1 + \left(20 \times 0.49 + \frac{1}{0.166}\right)^{\frac{1}{3}}\right]$$

$$= 0.0193$$

$$h = \left(f\frac{L}{d} + \Sigma K\right)\frac{v^2}{2g}$$

Substituting $f = 0.0193$, $\Sigma K = 1.9$, $L = 40$ m, $d = 0.092$ m, and $h = 1$ m:

$$1 = \left(\frac{0.0193 \times 40}{0.092} + 1.9\right) \times \frac{v^2}{19.62}$$

$$\therefore v = 1.38 \text{ m/s}$$

Trial 3

Take as the next trial value $v = 1.4$ m/s

$$0.01 = 1.4 \times \frac{\pi d^2}{4}$$

$$\therefore d = 0.095 \text{ m} = 95 \text{ mm}$$

It is clear that the friction factor will change very slightly at this velocity so $f = 0.0193$.

$$h = \left(f\frac{L}{d} + \Sigma K\right)\frac{v^2}{2g}$$

Substituting $f = 0.0193$, $\Sigma K = 1.9$, $L = 40$ m, $d = 0.095$ m, and $h = 1$ m:

$$1 = \left(\frac{0.0193 \times 40}{0.095} + 1.9\right) \times \frac{v^2}{19.62}$$

$$\therefore v = 1.4 \text{ m/s}$$

Therefore the solution is $v = 1.4$ m/s and the pipe size required is **95 mm**.

3.3 *PIPES IN SERIES*

Pipes are in series when a number of pipes of different diameters or materials (or both) are connected one after another so that the flow rate through each pipe is the same. Two pipes in series are shown in Figure 3.4.

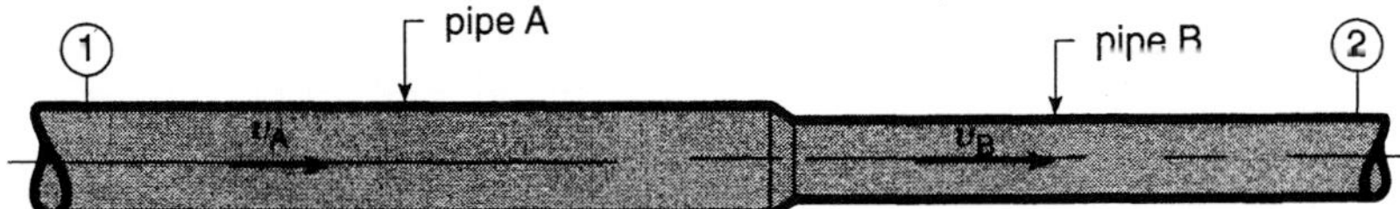

Fig 3.4 *Series pipes*

When pipes are in series, the head losses are additive, that is the total head loss is the sum of the head losses in each pipe. If there are two pipes, call one pipe A and the other B, then:

$$H_{L\,1-2} = H_{LA} + H_{LB} \qquad \qquad \dots\dots\dots\dots\dots\dots\dots\dots(1)$$

Because the flow rate through each pipe is the same, and assuming incompressible flow, by continuity:

$$\dot{V} = v_A A_A = v_B A_B \qquad \qquad \dots\dots\dots\dots\dots\dots\dots\dots(2)$$

These equations may be used to solve problems involving two pipes in series. In some cases (depending on the data given or known) a direct solution is possible by use of one or other of these equations, either separately or simultaneously. In other cases, a trial-and-error or iterative approach is necessary.

Notes

- If there are more than two pipes in series, these equations may be written in the more general form as:

$$H_{L\,1-2} = H_{LA} + H_{LB} + H_{LC} + \dots\dots\dots \qquad \qquad \dots\dots\dots\dots\dots\dots\dots\dots(1)$$

and

$$\dot{V} = v_A A_A = v_B A_B = v_C A_C = \dots\dots\dots \qquad \qquad \dots\dots\dots\dots\dots\dots\dots\dots(2)$$

- If the pipes in series are made of the same material, and there is not a very large difference in their diameter, the friction factor is approximately the same in all pipes.

Known flow rate

If the flow rate is known, a direct solution is possible because the friction factor and the head loss can be determined without a trial-and-error approach being necessary. This is demonstrated in Example 3.5.

Example 3.5

Two steel pipes are connected in series as shown in Figure 3.5. Pipe A is 100 mm in diameter and 1.5 km long. Pipe B is 80 mm in diameter and 1.0 km long. Water at 17°C is pumped through the pipes with the velocity in pipe A being 2 m/s. Determine the friction factor and head loss in each pipe and hence the total head loss. Minor losses (head loss due to the contraction) are negligible.

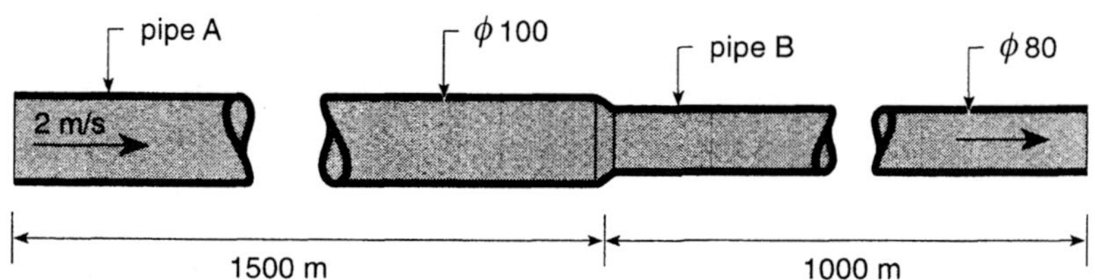

Fig. 3.5

Solution

From Appendix 13, interpolating: for water at 17°C, $\mu = 1.09 \times 10^{-3}$ Pas, $\rho = 999$ kg/m³. Also from Appendix 4, ε for steel $= 0.045$ mm.

Pipe A

$$\varepsilon_R = \frac{0.045}{100} = 0.45 \times 10^{-3}$$

$$\upsilon = 2 \text{ m/s}$$

$$Re = \frac{\upsilon d \rho}{\mu} = \frac{2 \times 0.1 \times 0.999 \times 10^3}{1.09 \times 10^{-3}} = 0.183 \times 10^6$$

(and flow is turbulent)

$$f = 0.0055 \left[1 + \left(20\,000\varepsilon_R + \frac{10^6}{Re} \right)^{\frac{1}{3}} \right]$$

$$= 0.0055 \left[1 + \left(20 \times 0.45 + \frac{1}{0.183} \right)^{\frac{1}{3}} \right]$$

$$= 0.0189$$

$$H_L = f \frac{L}{d} \frac{\upsilon^2}{2g}$$

$$= 0.0189 \times \frac{1500}{0.1} \times \frac{2^2}{19.62}$$

$$= \mathbf{57.8 \text{ m}}$$

Pipe B

$$\varepsilon_R = \frac{0.045}{80} = 0.5625 \times 10^{-3}$$

$$\upsilon = 2 \times \left(\frac{100}{80} \right)^2 = 3.125 \text{ m/s}$$

$$Re = \frac{v d \rho}{\mu} = \frac{3.125 \times 0.08 \times 0.999 \times 10^3}{1.09 \times 10^{-3}} = 0.229 \times 10^6$$

(and flow is turbulent)

$$f = 0.0055 \left[1 + \left(20\,000 \varepsilon_R + \frac{10^6}{Re} \right)^{\frac{1}{3}} \right]$$

$$= 0.0055 \left[1 + \left(20 \times 0.5625 + \frac{1}{0.236} \right)^{\frac{1}{3}} \right]$$

$$= 0.0192$$

$$H_L = f \frac{L}{d} \frac{v^2}{2g}$$

$$= 0.0192 \times \frac{1000}{0.08} \times \frac{3.125^2}{19.62}$$

$$= \mathbf{119.5\ m}$$

Total head loss $= 57.8 + 119.5 = 177.3$ m $= \mathbf{177\ m}$ (to 3 significant figures).

Note   From the above example, the difference in friction factor between the two pipes is not very great. The friction factor in Pipe A is 0.0189 and in Pipe B is 0.0192. The difference is about 2%, which is less than the likely order of accuracy in estimating head loss. That is, the assumption of the same friction factor leads to insignificant error if the pipes are made of the same material and the diameters are not greatly different.

3.4   *EQUIVALENT LENGTH – SERIES PIPES*

The solution to problems involving series pipes is often simplified by reducing the pipes to a single-diameter equivalent pipe. The equivalent length of a single pipe to replace a number of pipes of different diameters or materials in series, is *the length of single-diameter pipe that gives the same head loss for the same flow rate as the original series pipes.*

The equivalent length of several pipes in series may be calculated using the following formula (derived in Appendix 7):

$$\boxed{\ \frac{f_E L_E}{d_E{}^5} = \frac{f_A L_A}{d_A{}^5} + \frac{f_B L_B}{d_B{}^5} + \dots\ }$$

Equivalent length series pipes (3.1)

If the friction factor is assumed to be the same, Formula 3.1 reduces to:

$$\boxed{\ \frac{L_E}{d_E{}^5} = \frac{L_A}{d_A{}^5} + \frac{L_B}{d_B{}^5} + \dots\ }$$

Equivalent length series pipes
(same friction factor) (3.2)

The head loss can then be calculated using the Darcy formula:

$$H_L = f_E \frac{L_E}{d_E} \frac{v_E{}^2}{2g}$$

Notes

- In some cases it is convenient to use the equivalent length of the fittings in the system as outlined in Section 2.6. The equivalent length of fittings can be calculated using Formula 2.8, then the equivalent length of the fittings is added to the equivalent length of the pipes to obtain the total equivalent length.
- The diameter of the equivalent pipe can be chosen at any convenient value. However, to reduce calculation, it is best to make the diameter of the equivalent pipe the same as the diameter of one of the pipes in the original system. It is then not necessary to calculate the equivalent length of this pipe because the diameter has not been changed. This method is adopted in Example 3.6.
- In Formulas 3.1 and 3.2, the units of length must be the same for all pipes, and the units of diameter must also be the same for all pipes. However, it is not necessary for the units of diameter and length to be the same. For example, the diameter may be expressed in millimetre units, whereas the length may be in metre units.

Example 3.6

Solve Example 3.5 using:
(a) an equivalent pipe with diameter the same as Pipe A;
(b) an equivalent pipe with diameter the same as Pipe B.

Solution

From Example 3.5:

$$f_A = 0.0189, f_B = 0.0192, L_A = 1500 \text{ m}, L_B = 1000 \text{ m}, d_A = 100 \text{ mm},$$
$$d_B = 80 \text{ mm}$$

(a) In this case take $d_E = d_A = 100$ mm and $f_E = f_A = 0.0189$

For Pipe B,
$$\frac{f_E L_E}{d_E^5} = \frac{f_B L_B}{d_B^5}$$

Substituting:

$$\frac{0.0189 \times L_E}{0.1^5} = \frac{0.0192 \times 1000}{0.08^5}$$

$$\therefore L_E = 3100 \text{ m}$$

$$\therefore L_E \text{ (total)} = 3100 + 1500 = 4600 \text{ m}$$

$$H_L = f_E \frac{L_E}{d_E} \frac{v_E^2}{2g}$$

With $v_E = 2$ m/s from Example 3.5:

$$H_L = 0.0189 \times \frac{4600}{0.1} \times \frac{2^2}{19.62}$$

$$= \textbf{177 m} \text{ (as before)}$$

(b) In this case take $d_E = d_B = 80$ mm, and $f_E = f_B = 0.0192$

For Pipe A,
$$\frac{f_E L_E}{d_E^5} = \frac{f_A L_A}{d_A^5}$$

Substituting:

$$\frac{0.0192 \times L_{\mathrm{E}}}{0.08^5} = \frac{0.0189 \times 1500}{0.1^5}$$

$$\therefore L_{\mathrm{E}} = 483.8 \text{ m}$$

$$\therefore L_{\mathrm{E}} \text{ (total)} = 1000 + 483.8 = 1483.8 \text{ m}$$

$$H_{\mathrm{L}} = f_{\mathrm{E}} \frac{L_{\mathrm{E}}}{d_{\mathrm{E}}} \frac{v_{\mathrm{E}}^2}{2g}$$

With $v_{\mathrm{E}} = 3.125$ m/s from Example 3.5:

$$H_{\mathrm{L}} = 0.0192 \times \frac{1483.8}{0.08} \times \frac{3.125^2}{19.62}$$

$$= \mathbf{177 \ m} \text{ (as in (a))}$$

Unknown flow rate

The use of equivalent length in Example 3.6 did not simplify the calculation because the flow rate was given. In cases where the flow rate is unknown, the use of equivalent length simplifies the calculation of the flow rate through pipes in series. This is illustrated in Example 3.7.

Example 3.7

The two pipes given in Example 3.5 are used to drain a large tank containing water at 17°C as shown in Figure 3.6. Determine the flow rate. Neglect minor losses and the velocity head of the water leaving the pipe. Assume the friction factor is the same in both pipes and use a trial value of $f = 0.02$.

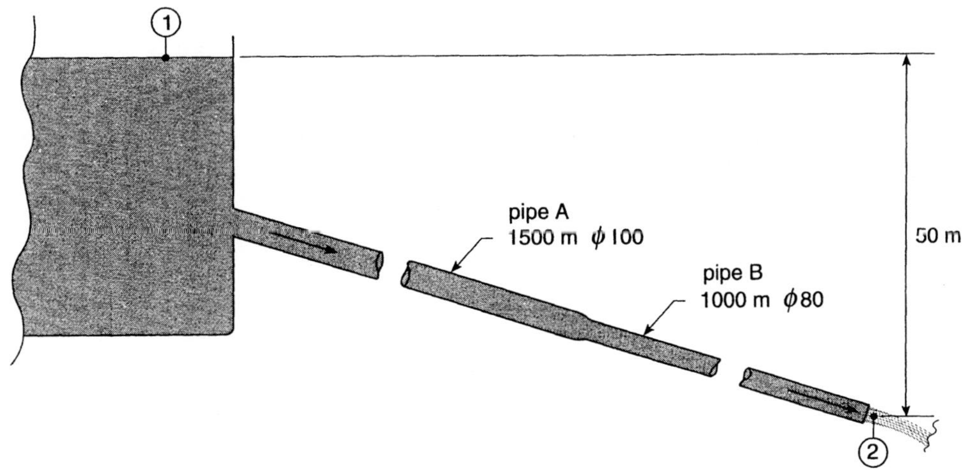

Fig. 3.6

Solution

In this case, Pipe A will be used as the basic pipe, and the equivalent length of Pipe B will be determined. Because the friction factor is assumed the same in both pipes, Formula 3.2 may be used:

$$\frac{L_E}{d_E{}^5} = \frac{L_B}{d_B{}^5} \qquad \text{or} \qquad L_E = \left(\frac{d_E}{d_B}\right)^5 L_B$$

Now $d_E = 100$ mm, $d_B = 80$ mm. and $L_B = 1000$ m

Substituting:

$$L_E = \left(\frac{100}{80}\right)^5 \times 1000 = 3052 \text{ m} \quad (1)$$

Therefore the total equivalent length of Pipe A and Pipe B in terms of A is $3052 + 1500 = 4552$ m. Neglecting the velocity head of the fluid at (2), Bernoulli's equation applied between (1) and (2) reduces to:

$$h = H_L = \left(f\,\frac{L}{d} + \Sigma K\right)\frac{v^2}{2g}$$

Also, if minor losses are neglected, $\Sigma K = 0$, hence:

$$h = f\,\frac{L}{d}\,\frac{v^2}{2g}$$

Substituting $h = 50$ m, $L = 4552$ m, $d = 0.1$ m, the equation becomes:

$$\frac{0.0216}{f} = v^2$$

Trial 1
$f = 0.02$, $\therefore v = 1.04$ m/s

$$Re = \frac{vd\rho}{\mu} = \frac{1.04 \times 0.1 \times 0.999 \times 10^3}{1.09 \times 10^{-3}} = 0.0953 \times 10^6$$

(and flow is turbulent)

$$\varepsilon_R = \frac{0.045}{100} = 0.45 \times 10^{-3}$$

$$f = 0.0055\left[1 + \left(20\,000\varepsilon_R + \frac{10^6}{Re}\right)^{\frac{1}{3}}\right]$$

$$= 0.0055 \times \left[1 + (9 + 10.49)^{\frac{1}{3}}\right]$$

$$= 0.0203 \text{ (and the initial trial value } f = 0.02 \text{ was very close)}$$

---

[1] This value is a little different to the value calculated in Example 3.6 of 3100 m. (difference about 1.4%) because now the friction factor has been assumed to have the same value in each pipe.

Trial 2

$f = 0.0203$, $\therefore v = 1.03$ m/s

$$Re = \frac{vd\rho}{\mu} = \frac{1.03 \times 0.1 \times 0.999 \times 10^3}{1.09 \times 10^{-3}} = 0.0944 \times 10^6$$

(and flow is turbulent)

$$\varepsilon_R = 0.45 \times 10^{-3}$$

$$f = 0.0055 \left[1 + \left(20\,000\varepsilon_R + \frac{10^6}{Re} \right)^{\frac{1}{3}} \right]$$

$$= 0.0055 \times \left[1 + (9 + 10.59)^{\frac{1}{3}} \right]$$

$$= 0.0203$$

Therefore the solution is $f = 0.0203$, and $v = 1.03$ m/s

$$\dot{V} = vA = 1.03 \times A(0.1) \text{ m}^3/\text{s} = \mathbf{8.09\ L/s}$$

Self-test problem 3.2

In the system shown in Figure 3.7, water with viscosity 1.1×10^{-6} m^2/s flows through steel pipes. There is a reducing elbow with K factor 0.9. Determine:
(a) the total equivalent length of the pipes if both pipes were 150 mm in diameter (assuming the same friction factor);
(b) the flow rate through the system (use an initial trial velocity of 10 m/s).

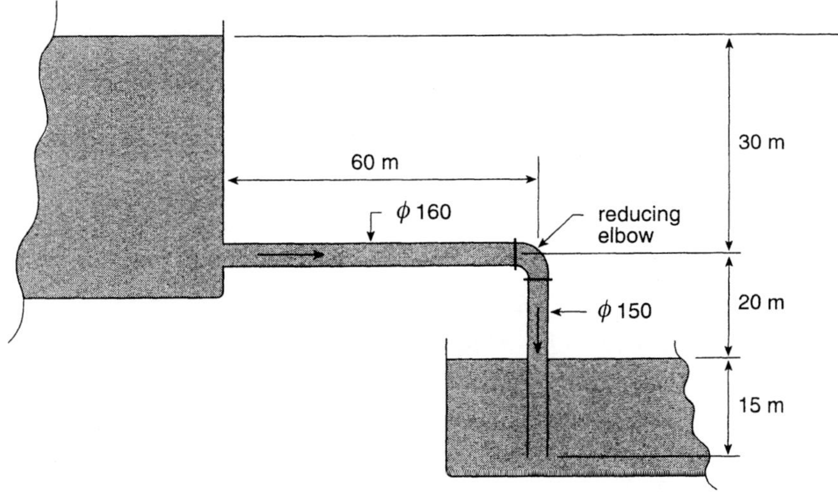

Fig. 3.7

3.5 *PIPES IN PARALLEL*

Pipes are in parallel when a number of pipes are connected in such a way that the flow is shared between the pipes. The pipes may be of the same or different diameters or materials. Some examples of pipes in parallel are shown in Figure 3.8 on page 66.

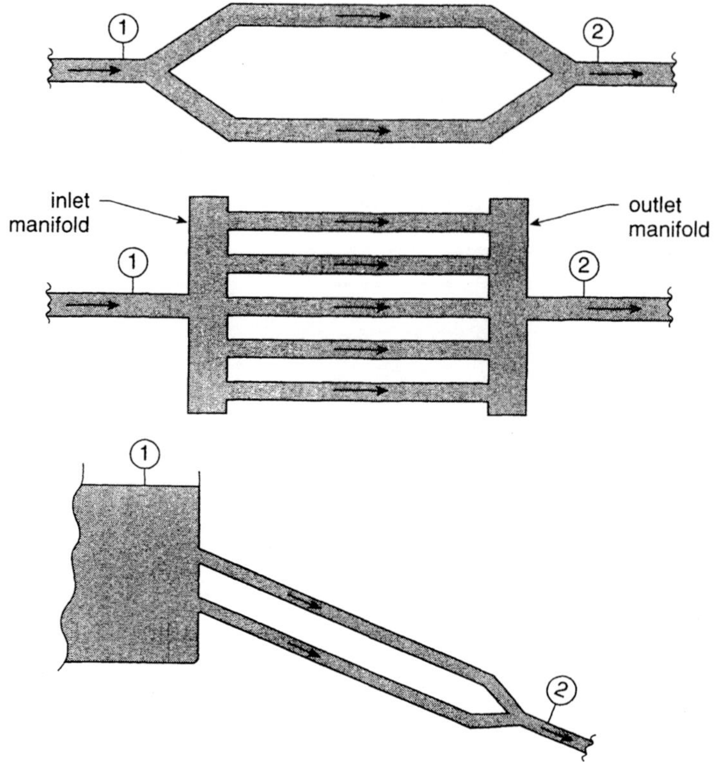

Fig. 3.8 *Parallel pipes*

When parallel pipes are connected at the upstream end (Point ①) to a common pipe, tank or reservoir, the pressure at this point is the same for all pipes. If the downstream pressure (Point ②) is also the same for all pipes, it follows that the head loss between ① and ② is the same in each pipe.

Consider the two pipes in parallel, and call one Pipe A and the other Pipe B. If the head loss is the same, then:

$$H_{L\,1-2} = H_{LA} = H_{LB} \qquad \text{.................................(1)}$$

Because the total flow rate is the sum of the flow rates in each pipe, by continuity:

$$\dot{V} = v_A A_A + v_B A_B \qquad \text{.................................(2)}$$

These two equations may be used to solve problems involving parallel pipes. In some cases (depending on the data given or known) a direct solution is possible by use of one or other of these equations, either separately or simultaneously. In other cases, a trial-and-error or iterative approach is necessary in order to obtain a solution.

Notes

- If there are more than two pipes in parallel, these equations may be written in the general form:

$$H_{L\,1-2} = H_{LA} = H_{LB} = H_{LC} = \text{..............} \qquad \text{...............................(1)}$$

and

$$\dot{V} = v_A A_A + v_B A_B + v_C A_C + \text{..............} \qquad \text{...............................(2)}$$

- If the pipes in parallel are made of the same material, and there is not a very large difference in their diameter, then the friction factor is approximately the same for both pipes. When this is the case, equation (1) for two pipes becomes:

$$f_A \frac{L_A}{d_A} \frac{v_A^2}{2g} = f_B \frac{L_B}{d_B} \frac{v_B^2}{2g}$$

But $f_A = f_B$

therefore:

$$\boxed{\frac{L_A v_A^2}{d_A} = \frac{L_B v_B^2}{d_B}}$$

Parallel pipes (same friction factor) (3.3)

Therefore if the friction factor is the same in the parallel pipes, the velocity (and flow rate) in each pipe is independent of the friction factor, and depends only on the length and diameter of the pipe.

Known flow rate or friction factor

If the flow rate in each pipe is known, or the friction factor in each pipe is given, a direct solution is possible because the head loss can be determined without a trial-and-error approach being necessary. This is demonstrated in Example 3.8.

Example 3.8

Water flows through the system shown in Figure 3.9 (opposite page) at a rate of 100 L/s. Determine:

(a) the flow rate in each of the branches B and C;

(b) the head loss between ① and ②.

Assume a friction factor of 0.02 in all pipes, and neglect minor losses. The diameters and lengths of the pipes are as listed below:

Pipe	Diameter (mm)	Length (m)
A	300	250
B	200	30
C	150	60
D	300	300

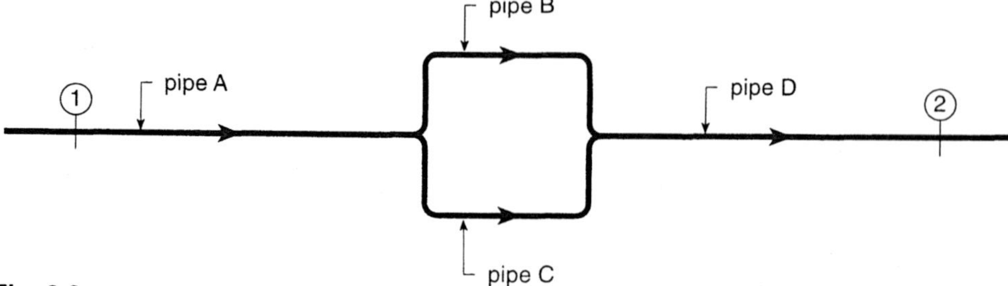

Fig. 3.9

Solution

(a) $v_A = v_D = \dfrac{\dot{V}}{A} = \dfrac{0.1}{A(0.3)} = 1.4147$ m/s

H_L (pipes A and D combined) $= f\dfrac{L}{d}\dfrac{v^2}{2g} = 0.02 \times \dfrac{550}{0.3} \times \dfrac{1.4147^2}{19.62} = 3.74$ m

H_L (pipe B) $= 0.02 \times \dfrac{30}{0.2} \times \dfrac{v_B{}^2}{19.62}$

H_L (pipe C) $= 0.02 \times \dfrac{60}{0.15} \times \dfrac{v_C{}^2}{19.62}$

Since the same friction factor has been assumed in both pipes, Formula 3.3 applies:

$$\dfrac{L_B v_B{}^2}{d_B} = \dfrac{L_C v_C{}^2}{d_C}$$

$$\therefore \dfrac{30 \times v_B{}^2}{0.2} = \dfrac{60 \times v_C{}^2}{0.15}$$

$$\therefore v_B{}^2 = 2.667 v_C{}^2$$

$$\therefore v_B = 1.633 v_C \qquad \qquad \text{.........................(1)}$$

Now $\qquad \dot{V}_B + \dot{V}_C = \dot{V} = 0.1$ m^3/s

$$\therefore v_B A_B + v_C A_C = 0.1$$

$$\therefore v_B A(0.2) + v_C A(0.15) = 0.1$$

$$\therefore 0.0314 v_B + 0.0177 v_C = 0.1 \qquad \qquad \text{.........................(2)}$$

Substituting $v_B = 1.633 v_C$ from Equation (1) into Equation (2):

$$0.0314 \times 1.633 v_C + 0.0177 v_C = 0.1$$

$$\therefore 0.069 v_C = 0.1$$

$$\therefore v_C = 1.45 \text{ m/s}$$

$$\therefore v_B = 1.633 \times 1.45 = 2.3676 \text{ m/s}$$

$$\therefore \dot{V}_B = 2.3676 \times 0.0314 = 0.0743 \text{ m}^3\text{/s} = \textbf{74.3 L/s}$$

$$\therefore \dot{V}_C = 1.45 \times 0.0177 = 0.0257 \text{ m}^3\text{/s} = \textbf{25.7 L/s}$$

Check: $\dot{V}_B + \dot{V}_C = 100$ L/s (Checks)

(b) H_L (pipe B) $= H_L$ (pipe C)

$$\therefore 0.02 \times \frac{30}{0.2} \times \frac{2.3676^2}{19.62} = 0.02 \times \frac{60}{0.15} \times \frac{1.45^2}{19.62} = 0.857 \text{ m}$$

$$\therefore H_L \text{ (total)} = 3.74 + 0.857 = \textbf{4.6 m}$$

3.6 *EQUIVALENT LENGTH—PARALLEL PIPES*

The solution to problems involving parallel pipes is often simplified by reducing the pipes to a single-diameter equivalent pipe. The equivalent length of a single pipe to replace a number of pipes in parallel is *the length of single-diameter pipe that gives the same head loss for the same flow rate as the original parallel pipes.*

The equivalent length of parallel pipes may be calculated using the following formula (derived in Appendix 8):

$$\left[\frac{d_E^{\,5}}{f_E L_E}\right]^{\frac{1}{2}} = \left[\frac{d_A^{\,5}}{f_A L_A}\right]^{\frac{1}{2}} + \left[\frac{d_B^{\,5}}{f_B L_B}\right]^{\frac{1}{2}} + \, \cdots$$

Equivalent length parallel pipes (3.4)

If the friction factor is assumed to be the same, Formula 3.4 reduces to:

$$\left[\frac{d_E^{\,5}}{L_E}\right]^{\frac{1}{2}} = \left[\frac{d_A^{\,5}}{L_A}\right]^{\frac{1}{2}} + \left[\frac{d_B^{\,5}}{L_B}\right]^{\frac{1}{2}} + \, \cdots$$

Equivalent length parallel pipes (same friction factor) (3.5)

The head loss can then be calculated using the Darcy formula:

$$H_L = f_E \frac{L_E}{d_E} \frac{v_E^{\,2}}{2g}$$

Notes

- In some cases it is convenient to use the equivalent length of the fittings in the system as outlined in Section 2.6. The equivalent length of fittings can be calculated using Formula 2.8, then the equivalent length of the fittings is added to the equivalent length of the pipes to obtain the total equivalent length.
- The diameter of the equivalent pipe can be chosen at any convenient value. However, to reduce calculation it is best to make the diameter of the equivalent pipe the same as the diameter of one of the pipes in the original system. It is then not necessary to calculate the equivalent length of this pipe because its diameter has not been changed. This method is adopted in Example 3.9.
- In Formulas 3.3, 3.4 and 3.5, the units of length must be the same for all pipes, and the units of diameter must also be the same for all pipes. However, it is not necessary for the units of diameter and length to be the same. For example, the diameter may be expressed in millimetre units, whereas the length may be in metre units.
- The use of Formulas 3.4 and 3.5 enables the total flow rate and head loss to be determined for pipes in parallel but does not enable the flow in *each* parallel pipe to be determined. If the friction factor is the same, Formula 3.3 may be used to calculate the flow rate in each pipe. Alternatively, if the head loss is known or calculated, the Darcy

formula may be used to calculate the velocity in each pipe because the head loss in each pipe is the same, and is equal to the overall head loss between all the parallel pipes.

Example 3.9

For the piping system given in Example 3.8, determine the head loss through the system using equivalent length for the parallel pipes.

Solution

Since the friction factor is assumed to have a constant value, which is the same in all pipes, Formula 3.5 may be used. The logical diameter to choose for the equivalent diameter is 300 mm since this is the diameter of pipes A and D. Formula 3.5 becomes:

$$\left[\frac{d_E^5}{L_E}\right]^{\frac{1}{2}} = \left[\frac{d_B^5}{L_B}\right]^{\frac{1}{2}} + \left[\frac{d_C^5}{L_C}\right]^{\frac{1}{2}}$$

where $d_E = 0.3$ m, $d_B = 0.2$ m, $d_C = 0.15$ m
         $L_E = ?$ $L_B = 30$ m, $L_C = 60$ m
Substituting:

$$\left[\frac{0.3^5}{L_E}\right]^{\frac{1}{2}} = \left[\frac{0.2^5}{30}\right]^{\frac{1}{2}} + \left[\frac{0.15^5}{60}\right]^{\frac{1}{2}}$$

$$\therefore \left[\frac{0.3^5}{L_E}\right]^{\frac{1}{2}} = 0.004\,39$$

$$\therefore \frac{0.3^5}{L_E} = 19.28 \times 10^{-6}$$

$$\therefore L_E = \frac{0.3^5}{19.28 \times 10^{-6}} = 126 \text{ m}$$

That is, the head loss in the parallel pipes B and C is the same as would occur in a 126 m length of 300 mm diameter pipe.

Therefore the total equivalent length of 300 mm diameter pipe is:
550 + 126 = 676 m.
The Darcy formula may now be applied to determine the total system head loss:

$$H_L = f\frac{L}{d}\frac{v^2}{2g} = 0.02 \times \frac{676}{0.3} \times \frac{1.4147^2}{19.62} = \textbf{4.6 m} \text{ (as before)}$$

Self-test problem 3.3

In Example 3.8 the friction factor was assumed to be 0.02 in all pipes. Now, do not assume the friction factor, but calculate it, based on the 300 mm diameter. You may still make the assumption that the friction factor is the same in each pipe. Determine:
(a) the friction factor;
(b) the flow rate in pipes B and C;
(c) the total head loss through the system.
All pipes are made of galvanised steel and the fluid is water with viscosity 1.1×10^{-6} m²/s.

Unknown flow rate or friction factor

When the flow rate or friction factor is not known (or given) a direct solution is not possible and a trial-and-error or iterative approach is required. If both the flow rate and friction factor are unknown, and the friction factor is not assumed to be the same in each pipe, extensive calculations are usually required (especially with more than two pipes in parallel). Such complex calculation procedures are best done by computer and several programs are available for solving pipe networks involving parallel or branching pipe systems. The solution of problems which require complex computations that do not introduce any new concepts or principles of fluid mechanics are outside the scope of this book.

Example 3.10

In the system shown in Figure 3.10, liquid with relative density 0.9 and viscosity 1×10^{-3} Pas flows from one tank to the other. The pipes are extruded and have diameters and lengths as follows:

Pipe	Diameter (mm)	Length (m)
A	50	70
B	60	50
C	80	40

When the globe valve is three-quarters open, determine:
(a) the flow rate through the system;
(b) the flow rate in pipes A and B.
Neglect minor losses, except for the entrance, exit and valve loss. Assume that the friction factor is the same in all pipes and use an initial trial value of $f = 0.02$.

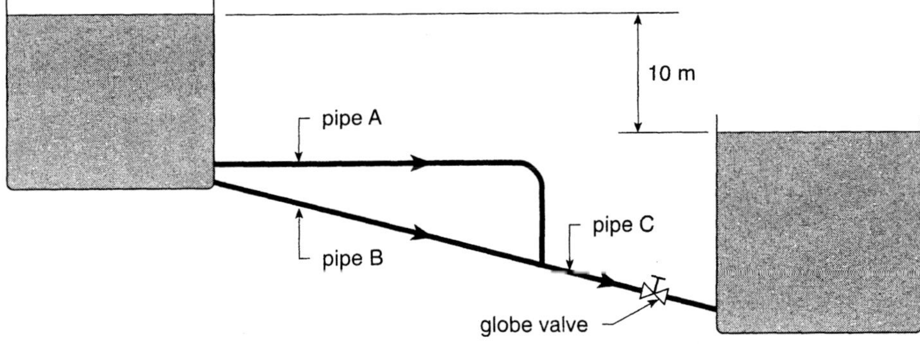

Fig. 3.10

Solution

(a) First determine the equivalent length of Pipes A and B in terms of C.
   Because the friction factor is assumed to be the same, use Formula 3.5:

$$\left[\frac{d_E^5}{L_E}\right]^{\frac{1}{2}} = \left[\frac{d_A^5}{L_A}\right]^{\frac{1}{2}} + \left[\frac{d_B^5}{L_B}\right]^{\frac{1}{2}}$$

where $d_E = 80$ mm, $d_A = 50$ mm, $d_B = 60$ mm

              $L_E = ?$ $L_A = 70$ m $L_B = 50$ m

Substituting:

$$\left[\frac{80^5}{L_E}\right]^{\frac{1}{2}} = \left[\frac{50^5}{70}\right]^{\frac{1}{2}} + \left[\frac{60^5}{50}\right]^{\frac{1}{2}}$$

$$\therefore \frac{80^5}{L_E} = 36.68 \times 10^6$$

$$\therefore L_E = \frac{80^5}{36.68 \times 10^6} = 89.33 \text{ m}$$

$$\therefore L_E \text{ (total)} = 89.33 + 40 = 129.33 \text{ m}$$

$$H_L = \left(f\frac{L}{d} + \Sigma K\right)\frac{v^2}{2g} = 10 \text{ m} \quad\quad\quad\quad \text{..................(1)}$$

$$\Sigma K = 0.5 \text{ (entrance)} + 1.0 \text{ (exit)} + 8.0 \ (\tfrac{3}{4} \text{ open globe valve)} = 9.5$$

Trial 1

$f = 0.02$, and substituting in Equation (1):

$$\left(0.02 \times \frac{129.33}{0.08} + 9.5\right)\frac{v^2}{19.62} = 10$$

$$\therefore v = 2.17 \text{ m/s}$$

Trial 2

$v = 2.17$ m/s

$$Re = \frac{v d \rho}{\mu} = \frac{2.17 \times 0.08 \times 900}{1 \times 10^{-3}} = 0.156 \times 10^6$$

$$\varepsilon_R = 0 \text{ (extruded tube)}$$

$$f = 0.0055\left[1 + \left(20\,000\varepsilon_R + \frac{10^6}{Re}\right)^{\frac{1}{3}}\right]$$

$$= 0.0055\left[1 + \left(0 + \frac{1}{0.156}\right)^{\frac{1}{3}}\right]$$

$$= 0.0157$$

Substituting in Equation (1):

$$\left(0.0157 \times \frac{129.33}{0.08} + 9.5\right)\frac{v^2}{19.62} = 10$$

$$\therefore v = 2.37 \text{ m/s}$$

Trial 3

$v = 2.37$ m/s

$$Re = \frac{vd\rho}{\mu} = \frac{2.37 \times 0.08 \times 900}{1 \times 10^{-3}} = 0.171 \times 10^6$$

$$f = 0.0055 \left[1 + \left(20\,000\varepsilon_R + \frac{10^6}{Re}\right)^{\frac{1}{3}}\right]$$

$$= 0.0055 \left[1 + \left(0 + \frac{1}{0.171}\right)^{\frac{1}{3}}\right]$$

$$= 0.0154$$

Substituting in Equation (1):

$$\left(0.0154 \times \frac{129.33}{0.08} + 9.5\right)\frac{v^2}{19.62} = 10$$

$$\therefore v = 2.39 \text{ m/s}$$

Clearly there is little to be gained from further trials, so $v = 2.39$ m/s

$$\dot{V} = vA = 2.39 \times A(0.08) \text{ m}^3/\text{s} = 12.01 \text{ L/s, say, } \mathbf{12 \text{ L/s}}$$

(b)
$$\dot{V} = v_C A_C = v_A A_A + v_B A_B$$
$$\therefore v_C d_C^2 = v_A d_A^2 + v_B d_B^2$$
$$\therefore 2.39 \times 80^2 = v_A \times 50^2 + v_B \times 60^2$$
$$\therefore 15\,296 = v_A \times 2500 + v_B \times 3600 \qquad \dots\dots\dots\dots (2)$$

From Formula 3.3,

$$\frac{L_A v_A^2}{d_A} = \frac{L_B v_B^2}{d_B}$$
$$\therefore \frac{70 v_A^2}{50} = \frac{50 v_B^2}{60}$$
$$\therefore 1.4 v_A^2 = 0.833 v_B^2$$
$$\therefore v_A = 0.7715 v_B$$

Substituting in Equation (2):

$$15\,296 = 0.7715 v_B \times 2500 + v_B \times 3600$$
$$\therefore 15\,296 = v_B \times 1929 + v_B \times 3600 = 5529 v_B$$
$$\therefore v_B = 2.766 \text{ m/s}$$
$$\therefore v_A = 2.134 \text{ m/s}$$

$$\dot{V}_A = v_A A_A = 2.134 \times A(0.05) \text{ m}^3/\text{s} = \mathbf{4.19 \text{ L/s}}$$
$$\dot{V}_B = v_B A_B = 2.766 \times A(0.06) \text{ m}^3/\text{s} = \mathbf{7.82 \text{ L/s}}$$

Check: $\dot{V}_A + \dot{V}_B = 12.01$ L/s (Checks)

Alternative solution using the Darcy formula

$v_C = 2.39$ m/s and $f = 0.0154$

Using the Darcy formula:

$$H_{LC} = \left(f \frac{L_C}{d_C} + \Sigma K \right) \frac{v_C^2}{2g}$$

In Pipe C, $\Sigma K = 9.0$ (same as ΣK before, but excluding the entrance loss)

$$L_C = 40 \text{ m and } d_C = 80 \text{ mm}$$

Substituting:

$$H_{LC} = \left(0.0154 \times \frac{40}{0.08} + 9.0 \right) \times \frac{2.39^2}{19.62} = 4.862 \text{ m}$$

$$\therefore H_{LA} = H_{LB} = 10 - 4.862 = 5.138 \text{ m}$$

For Pipe A:

$$H_{LA} = \left(f \frac{L_A}{d_A} + \Sigma K \right) \frac{v_A^2}{2g}$$

$$\therefore 5.138 = \left(0.0154 \times \frac{70}{0.05} + 0.5 \right) \times \frac{v_A^2}{19.62}$$

$\therefore v_A = 2.138$ m/s (which is within 0.2% of the value 2.134 m/s obtained previously)

For Pipe B:

$$H_{LB} = \left(f \frac{L_B}{d_B} + \Sigma K \right) \frac{v_B^2}{2g}$$

$$\therefore 5.138 = \left(0.0154 \times \frac{50}{0.06} + 0.5 \right) \times \frac{v_B^2}{19.62}$$

$\therefore v_B = 2.751$ m/s (which is within 0.5% of the value 2.766 m/s obtained previously)

 ## *Self-test problem 3.4*

Water with viscosity 1.15×10^{-6} m²/s flows from one reservoir to another through the piping system shown in Figure 3.11 opposite. All pipes have the same diameter and are made of concrete with roughness of 0.2 mm. Determine the minimum required diameter of the pipes if the flow rate is to be at least 500 L/s. Assume the same friction factor in all pipes, and neglect minor losses. Use an initial trial value of $v = 1$ m/s.

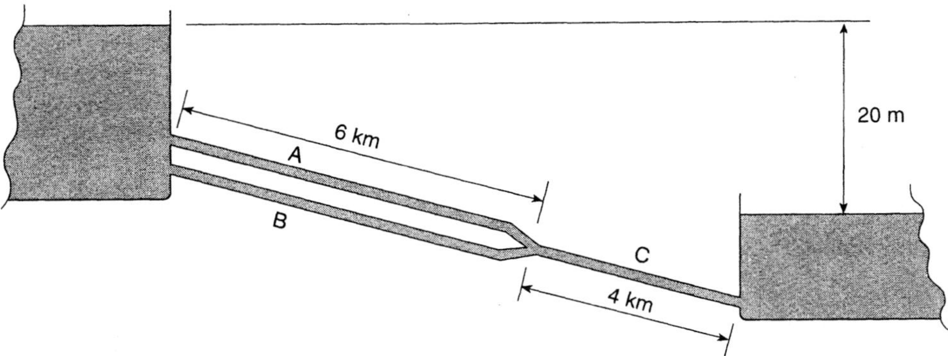

Fig. 3.11

Summary

A common example of the free flow of a fluid occurs when the fluid flows under the influence of gravity out of a storage tank or reservoir. In this case if flow losses are neglected, the Bernoulli equation reduces to: $v = \sqrt{2gh}$, (the Torricelli equation) where h is the height of the free surface of the fluid above the outlet point. When fluid flows from a non-vented tank under pressure, the pressure head in the tank must be included, so the equation becomes: $v = \sqrt{2gH}$, where H is the total head of the fluid in the tank. When the head loss is included, the Torricelli equation becomes: $v = \sqrt{2g(H - H_L)}$.

If fluid flows from one tank to another, the velocity head at outlet is zero and the Bernoulli equation reduces to: $H = H_L$, that is, all the static fluid head is dissipated as head loss.

When applying any of these equations the solution is straightforward when the head loss is known (or given). When the head loss is not known (or given), the solution becomes more complex and a trial-and-error (or iterative) approach may be needed.

It will be found that most iterative solutions converge rapidly and, for any reasonable initial trial values, a solution consistent with the required overall order of accuracy can usually be obtained within two or three iterations. Even for a pure guess of the initial trial values it will be found that only one or two more iterations will be required in order to produce the same order of accuracy of the solution.

If the friction factor is known (or given) the solution is more direct than when the friction factor is not given. If the friction factor is not known (or given) an initial trial value of $f = 0.02$ is usually a good starting point.

In many cases (particularly when there are long lengths of pipe involved) minor losses may be negligible compared to the frictional loss in the pipe. If this is not the case, the K factors for fittings (including entrance and outlet) need to be included when determining the head loss.

In some cases there may be a multiplicity of different pipes connected either in series or parallel, or both. When pipes are in series, the flow rate through each pipe is the same, and the total head loss is the sum of the head losses occurring in each pipe. When pipes are in parallel, the head loss is the same in each pipe, and the flow rate is the sum of the flow rates in each of the pipes.

The solution to series and parallel piping systems is often simplified by using an equivalent pipe. The length of the equivalent pipe is such that the head loss in the equivalent pipe is the same as the head loss that occurs in the original system (for the same flow rate of fluid). Formulas 3.1 and 3.2 are given in the text (and derived in Appendix 7) for calculating the equivalent length of pipes in series. Formulas 3.4 and 3.5 apply for pipes in parallel (derived in Appendix 8).

In most cases, little accuracy is lost by assuming the friction factor is the same in each of the series or parallel pipes, and this assumption greatly simplifies the calculation. Formulas 3.2 and 3.4 are based on this assumption.

In order to calculate the flow rate in individual branches of pipes in parallel, Formula 3.3 (derived in the text) may be used. Alternatively, if the head loss is known or calculated, the Darcy formula may be used to calculate the velocity in each pipe because the head loss in each pipe is the same, and is equal to the overall head loss between all the parallel pipes.

If the friction factor is not assumed to be the same in all pipes in parallel the solution may require considerable computation, which is best done by computer. Problems requiring extensive computation for a solution have not been included in this book because they do not aid in the understanding of the principles involved in the flow of fluids through pipe systems.

 # *Problems*

Notes
- Assume that the density of water is 1000 kg/m^3 unless stated otherwise.
- When surface roughness is not given, use Appendix 4.
- When K factors are not given. use Appendix 6.
- When pipes are in series or parallel. assume the same friction factor in all pipes, unless otherwise stated.

3.1 In the system shown in Figure P3.1 opposite, water flows at a steady rate of 10 L/s. The pipe is 65 mm diameter galvanised steel. The water viscosity is 0.9×10^{-6} m^2/s and the density is 996 kg/m^3. Minor losses may be neglected but the velocity head of the water in the pipe should be included. Determine:
   (a) the reading shown by the pressure gauge;
   (b) the length L of pipe below the gauge.
   (a) 21.2 kPa (b) 25.8 m

3.2 Water at 60°C drains from the tank through the system illustrated in Figure P3.2 opposite. Use Appendix 13 for the density and viscosity of water at this temperature, and determine:
   (a) the flow rate if losses are neglected:
   (b) the flow rate if losses are included. Two trials only are required, using an initial trial value of 2 m/s for the velocity v of the water in the tube.
   (a) 0.4165 L/s (b) Trial 1, 0.137 L/s ($f = 0.022$, $v = 1.742$ m/s);
   Trial 2. 0.136 L/s ($f = 0.0227$. $v = 1.73$ m/s)

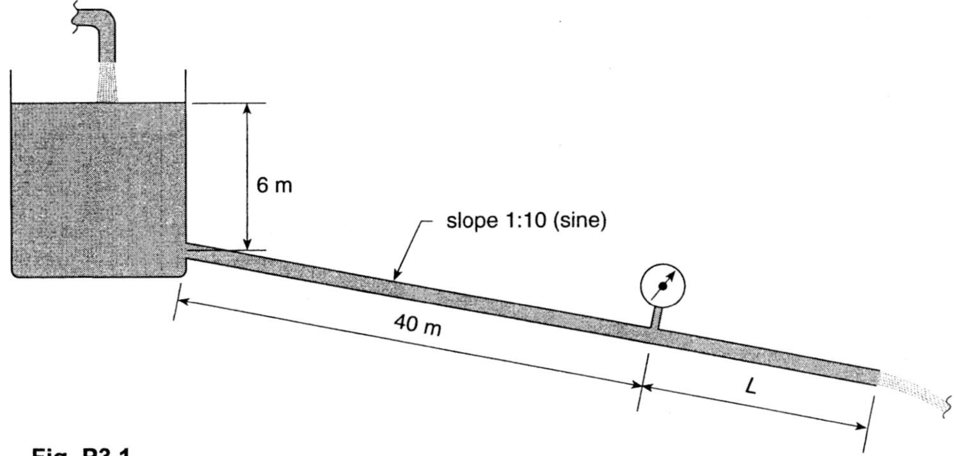

Fig. P3.1

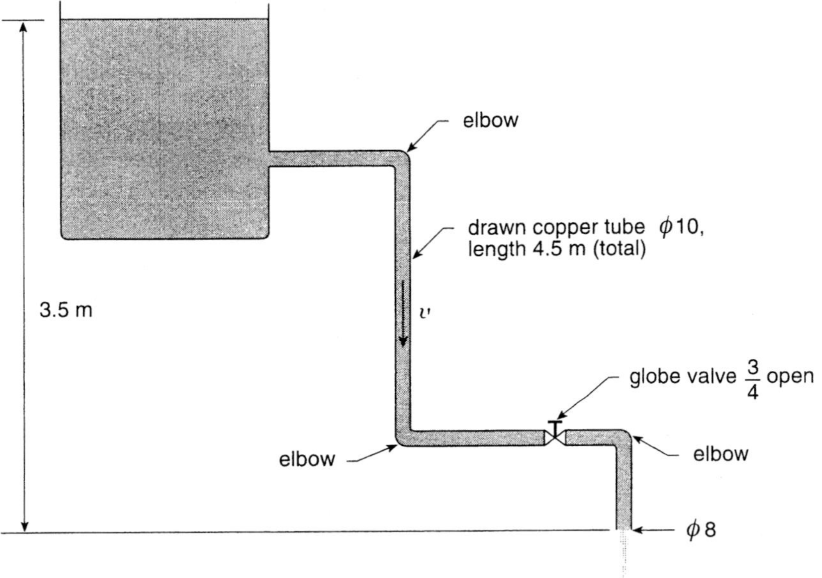

Fig. P3.2

3.3 For the system shown in Figure P3.3 on page 78, water with viscosity 1.1×10^{-6} m^2/s flows through the steel pipes at a rate of 180 L/s. There is a reducing elbow with K factor 0.9. Determine:

(a) the friction factor in the 150 mm diameter pipe;
(b) the equivalent length of 150 mm diameter pipe;
(c) the diameter of the horizontal pipe (to the nearest mm).

    (a) 0.0159 (b) 66.65 m (c) 170 mm

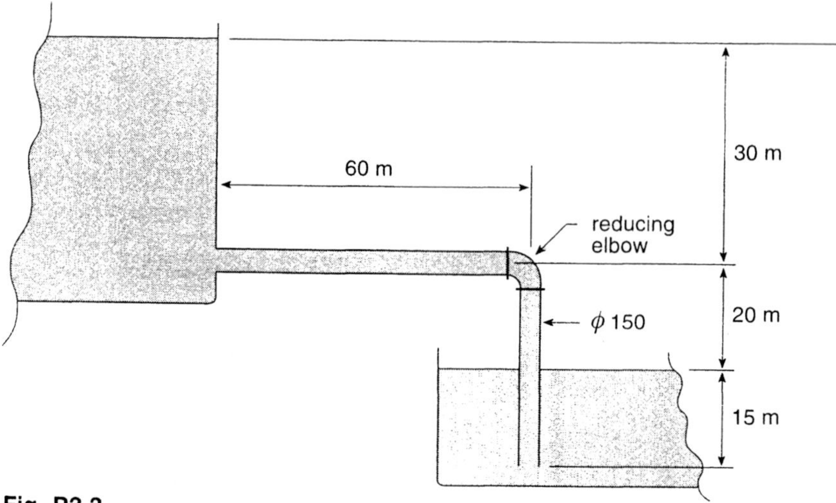

Fig. P3.3

3.4 In an airconditioning application as illustrated in Figure P3.4, return air with viscosity 18×10^{-6} Pas and specific volume 0.81 m³/kg flows through a square galvanised steel duct of side dimension 300 mm and total length 30 m. The pressure in the duct at ① is 20 mm (water gauge) and at ② is –15 mm (water gauge). The sum of the K factors in the return air duct is 3.5. Determine for the return air duct:

(a) the head loss;

(b) the average air velocity;

(c) the air flow rate.

When solving (b), use an initial trial value $f = 0.02$.

(a) 28.35 m (b) 10.15 m/s (c) 0.913 m³/s

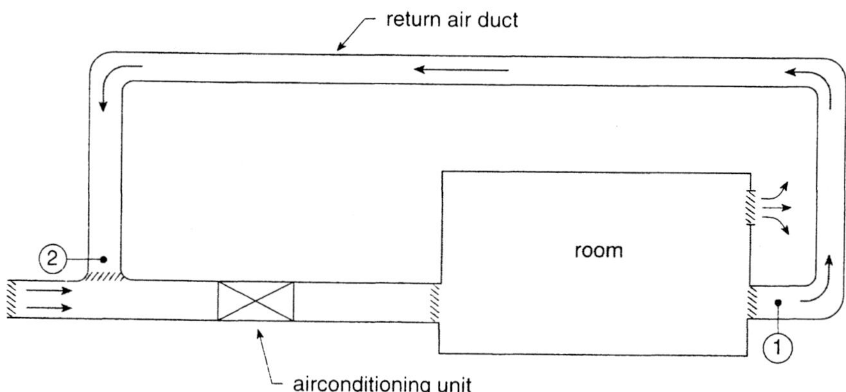

Fig. P3.4

3.5 As shown in Figure P3.5 opposite, water with viscosity 1.3×10^{-6} m²/s flows through two steel pipes joined in series. The flow rate is 10 L/s and the pressure reading shown by the pressure gauge at ① is 150 kPa. Neglect the flow loss due to the enlargement. Determine:

(a) the friction factor for pipe A;

(b) the friction factor for pipe B;

(c) the head loss in pipe A;
(d) the head loss in pipe B;
(e) the total head loss;
(f) the reading shown by the pressure gauge at ②.
  (a) 0.0202 (b) 0.0207 (c) 1.67 m (d) 0.45 m (e) 2.12 m (f) 129 kPa

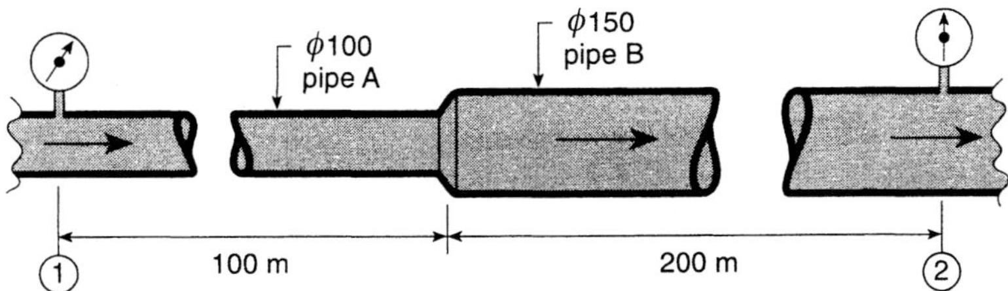

Fig. P3.5

3.6 (a) For the system shown in Figure P3.5, if the 150 mm diameter pipe were replaced by a 100 mm diameter pipe, what total length of 100 mm diameter pipe would give the same head loss between points ① and ② as the original system? Obtain an exact answer using the Darcy formula and an approximate answer assuming the same friction factor.
(b) For the system shown in Figure P3.5, if the 100 and 150 mm diameter pipes were replaced by a 125 mm diameter pipe, what total length of 125 mm diameter pipe would give the same head loss between points ① and ② as the original system? Obtain an exact answer using the Darcy formula and an approximate answer assuming the same friction factor.
  (a) 127 m, 126.3 m (b) 384 m, 385.6 m

3.7 Fluid flows through the system shown in Figure P3.7. Assuming a constant friction factor of 0.02 and neglecting minor losses, determine:
(a) the equivalent length of 250 mm diameter pipe for the two parallel pipes A and B;
(b) the head loss between points ① and ②;
(c) the velocity in pipes A and B.
  (a) 128 m (b) 4.69 m (c) 2.938 m/s, 3.035 m/s

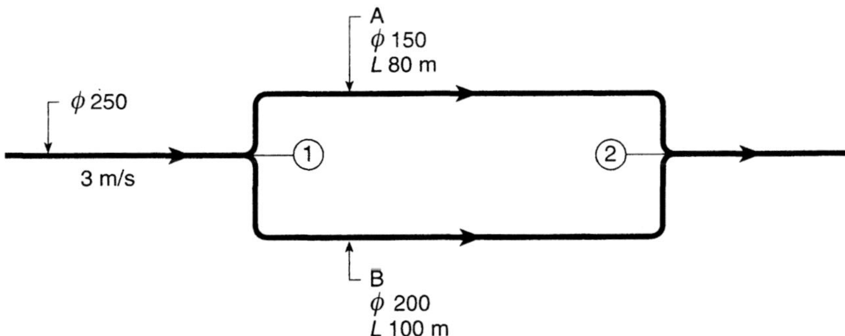

Fig. P3.7

3.8 As shown in Figure P3.8 on page 80, kerosene with relative density 0.78 and viscosity 2×10^{-3} Pas flows from an upper storage tank under a pressure of 50 kPa (gauge) to a lower vented storage tank. Minor losses are negligible, and the pipes are extruded and can be considered to be smooth. Determine:

  (a) the head loss in the system;
  (b) the total equivalent length of 25 mm diameter pipe;
  (c) the flow rate (using an initial trial value f = 0.02 and performing 3 trials).
      (a) 26.53 m (b) 131 m (c) 0.959 L/s, Trial 1, f = 0.02, v = 2.23 m/s;
      Trial 2, f = 0.0252, v = 1.98 m/s; Trial 3, f = 0.026, v = 1.95 m/s

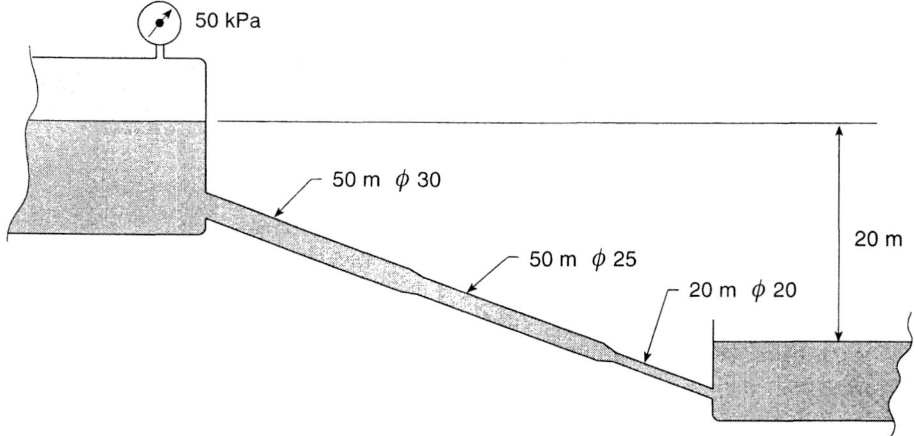

Fig. P3.8

3.9 As shown in Figure P3.9, liquid with viscosity 2.5×10^{-6} m²/s drains from a tank. The valve is a fully open globe valve and the pipe is made of 30 mm diameter drawn copper.
  (a) Determine the friction factor and velocity. Use an initial trial value of f = 0.02 and perform three trials.
  (b) Use the third trial value from (a) to determine the flow rate.
      (a) Trial 1, f = 0.02, v = 1.372 m/s; Trial 2, f = 0.0272, v = 1.237 m/s;
        Trial 3, f = 0.028, v = 1.225 m/s (b) 0.866 L/s

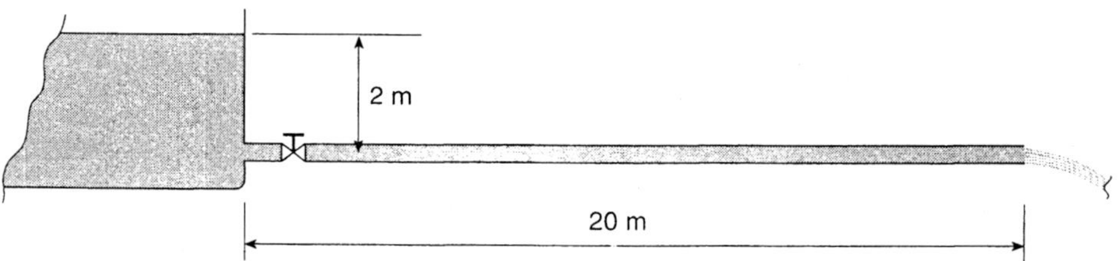

Fig. P3.9

3.10 For the system shown in Figure P3.9, the flow rate is to be 3 L/s.
  (a) Determine the pipe diameter required. Assume a constant friction factor of f = 0.025.
  (b) Calculate the friction factor for the pipe diameter obtained in (a) and hence check if the assumed value was reasonable.
      (a) 50 mm (b) 0.0232

3.11 Sea water with relative density 1.04 and viscosity 1.2×10^{-3} Pas flows through the system shown in Figure P3.11 (opposite) at a rate of 4 L/s. The pressure at ① is 100 kPa (gauge). Determine:
  (a) the total equivalent length in terms of pipe D;
  (b) the pressure shown by the pressure gauge at ②;

(c) the flow rate in each of the pipes B and C.

Assume a constant friction factor of 0.02 in all pipes, and neglect minor losses. The pipes are made of drawn tube and their diameters and lengths are as listed in the table:

Pipe	Diameter (mm)	Length (m)
A	40	20
B	40	20
C	50	30
D	50	10

(a) 81.4 m (b) 29.7 kPa (c) $\dot{V}_B$ = 1.65 L/s, $\dot{V}_C$ = 2.35 L/s

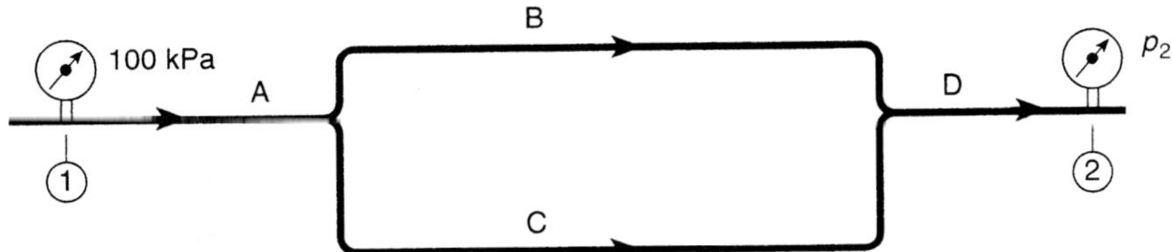

Fig. P3.11

3.12 For the system given in Problem 3.11, calculate:
  (a) the friction factor in the equivalent pipe (pipe D);
  (b) the pressure shown by the gauge at ②, using the new calculated value of the friction factor (not the previously assumed value).
  (a) 0.0181 (b) 36.5 kPa

3.13 In the system shown in Figure P3.13, valve A is closed and valve B is fully open. Water with viscosity 1.15×10^{-6} m²/s is pumped at a rate of 375 L/s from the lower reservoir to the upper reservoir. Galvanised steel pipe is used throughout. Minor losses are negligible.

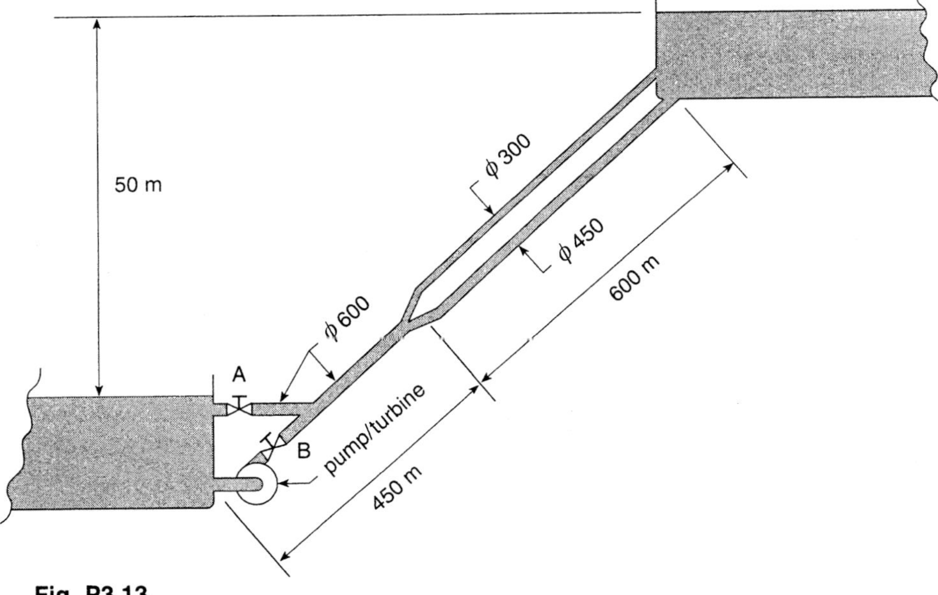

Fig. P3.13

Determine:
(a) the total equivalent length of 600 mm diameter pipe;
(b) the friction factor;
(c) the head loss;
(d) the total pump head;
(e) the fluid power provided by the pump;
(f) the flow rate in the 300 mm and 450 mm diameter branches.

(a) 1811 m (b) 0.0157 (c) 4.26 m (d) 54.26 m (e) 200 kW (f) 100 L/s, 275 L/s

3.14 In the system shown in Figure P3.13, valve B is now closed, and valve A is fully open. The pump is shut off, so the water flows back down from the upper reservoir to the lower one.

(a) Determine the friction factor and velocity in the 600 mm diameter pipe. Perform three trials using an initial trial value $f = 0.02$.
(b) Using the values obtained from Trial 3, determine the flow rate.

(a) Trial 1, $f = 0.02$, $v = 4.03$ m/s; Trial 2, $f = 0.0152$, $v = 4.625$ m/s; Trial 3, $f = 0.0152$, $v = 4.63$ m/s (b) 1.31 m^3/s

3.15 In the system shown in Figure P3.13, valve A is now closed and valve B is fully open. Water flows back down from the upper to the lower reservoir and drives the turbine. The fluid power transferred to the turbine is 100 kW. Assuming a constant friction factor $f = 0.015$, determine the velocity in the 600 mm diameter pipe, accurate to three significant figures and hence the flow rate. Use an initial trial value $v = 1$ m/s.

0.740 m/s, 209 L/s

3.16 In the system shown in Figure P3.16, water with viscosity 1×10^{-6} m^2/s flows through the system. The pressure at ① is 50 kPa, and at ② is 20 kPa. The pipes are smooth and minor losses are negligible. Determine:
(a) the equivalent length in terms of pipe A;
(b) the friction factor in pipe A, accurate to three significant figures, using an initial trial value $f = 0.02$;
(c) the total flow rate;
(d) the flow rate in pipes B, C and D.

(a) 25.1 m (b) Trial 1, $f = 0.02$, $v = 2.44$ m/s; Trial 2, $f = 0.0166$, $v = 2.68$ m/s; Trial 3, $f = 0.0162$, $v = 2.71$ m/s; Trial 4, $f = 0.0162$, $v = 2.72$ m/s; Trial 5, $f = 0.0162$, $v = 2.72$ m/s (solution) (c) 5.33 L/s
(d) $\dot{V}_B = 1.26$ L/s, $\dot{V}_C = 2.69$ L/s, $\dot{V}_D = 1.38$ L/s

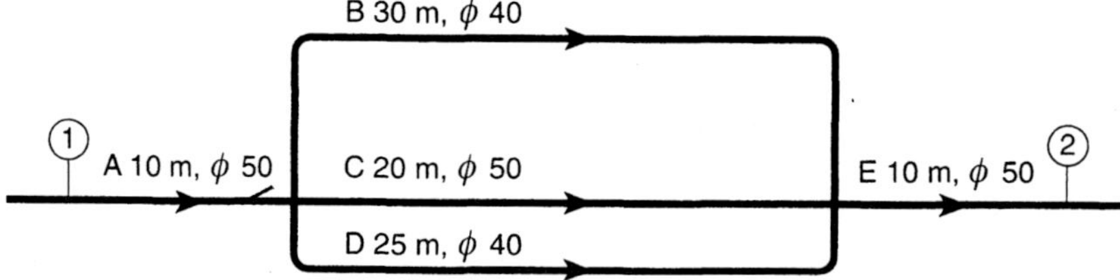

Fig. P3.16

CHAPTER 4

Channel flow

Objectives

On completion of this chapter you should be able to:

- recognise the distinguishing features that apply to channel flow;
- state the meaning of the various terms used in channel flow;
- solve channel flow problems using the Bernoulli equation and the Darcy formula;
- solve channel flow problems using the Chezy formula;
- solve channel flow problems using the Manning formula;
- recognise optimum channel shapes for both constant and variable flow rates.

Introduction

Channel flow occurs when a liquid flows down an inclined duct or pipe with a free upper surface. That is, in channel flow the upper surface of the liquid is not in contact with the duct or pipe wall. This occurs when liquids flow in partially full pipes or ducts, or in open channels, gutters, drains or culverts. Because this is a common situation, it is important that engineers understand it.

The principles of fluid flow in pipes running full also apply to channel flow, except that there are no pressure forces driving the fluid. The fluid flows as a result of the force of gravity acting down the inclined plane (the slope of the channel). Hence channel flow occurs only when the downstream end of the channel is at a lower elevation than the upstream end. Channel flow usually involves the flow of water, or water-based mixtures or slurries (such as sewage). Practical situations involving non-water-based fluids in channel flow are rare.

Some special formulas have been developed for channel flow, the most common ones being the Chezy and Manning formulas. In this chapter these formulas are presented and used. However, channel flow problems can also be solved using basic principles of fluid mechanics applicable to the flow of fluids in pipes or ducts, in particular the Bernoulli and Darcy equations. This method is presented first in this chapter.

4.1 *CHANNEL FLOW — BASIC CONCEPTS*

Channel flow occurs when:

- a liquid flows in a sloped open channel;
- a liquid flows partially full in a sloped enclosed pipe or duct.

These two variations are illustrated in Figure 4.1.

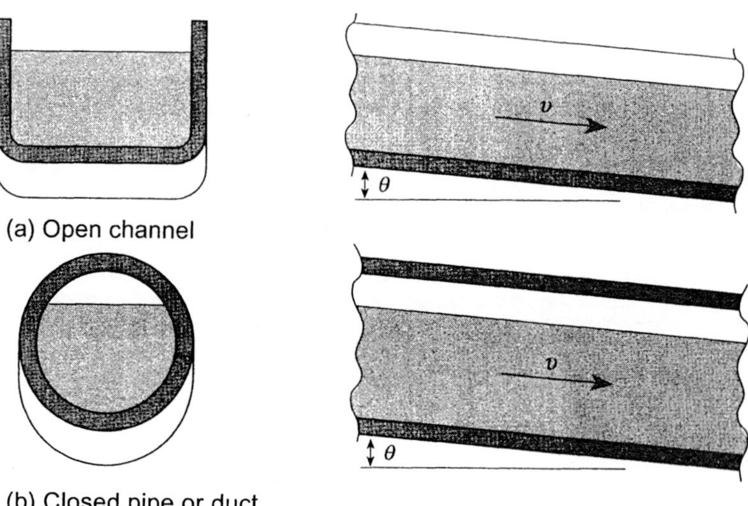

Fig. 4.1   *Channel flow*

Because the flow is driven purely by the force of gravity, it is analogous to the rolling of a ball down an inclined plane. This is illustrated in Figure 4.2.

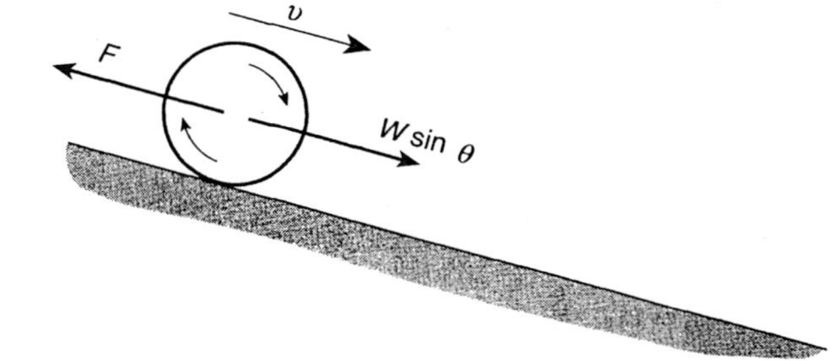

Fig. 4.2   *Solid equivalent to channel flow with a liquid*

Initially the ball accelerates down the plane until it reaches terminal velocity. At this velocity the friction force F acting up the plane balances the $W \sin \theta$ force exerted by gravity down the plane. As will be seen, the same situation occurs with fluids in channel flow.

Shape of the channel

There are many possible shapes of channels; the most common sections are illustrated in Figure 4.3 (opposite). They are: (a) circular, (b) oval, (c) rectangular (or square), (d) vee (or triangular), (e) trapezoidal, and (f) compound. Optimum sections for channel flow are discussed later in this chapter.

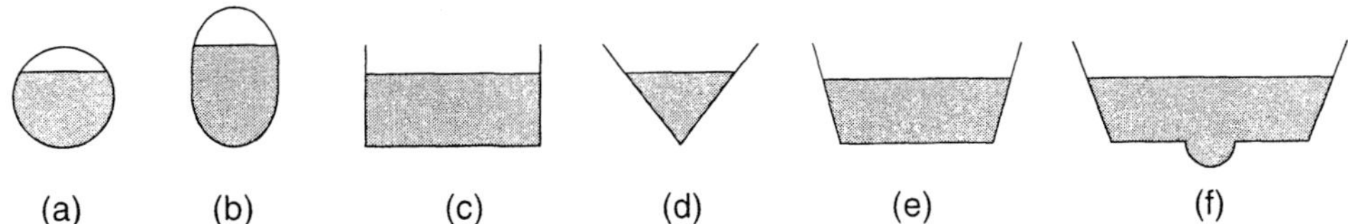

(a)   (b)   (c)   (d)   (e)   (f)

Fig. 4.3 *Typical channel sections*

Slope of the channel

The slope of the channel is the inclination of the channel with the horizontal. As illustrated in Figure 4.1, one method of defining the slope of the channel is by the angle θ made by the channel with the horizontal. However, many channels have a small slope and the angle θ is small, so it is often more convenient to define the slope of the channel as the vertical fall divided by the run (length along the slope). In this book the symbol S is used for slope thus defined, and $S = \sin \theta$.[1] The meaning of a channel with a slope of 1 in 1000, or 1/1000 is illustrated in Figure 4.4. For this slope $S = 0.001$ and the angle $\theta = 0.0573°$, which is a small angle, but sufficient to ensure good flow with a low viscosity liquid such as water.

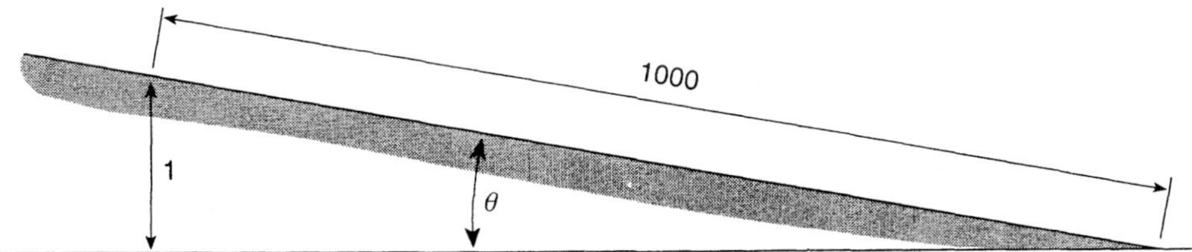

Fig. 4.4 *Channel slope (not to scale)*

In some cases the slope of the channel may change along its length but in this book it is assumed than all channels have constant slope.

Steady flow

Consider a liquid flowing out of a reservoir or tank through a sloped open channel to another reservoir or tank below it, as shown in Figure 4.5.

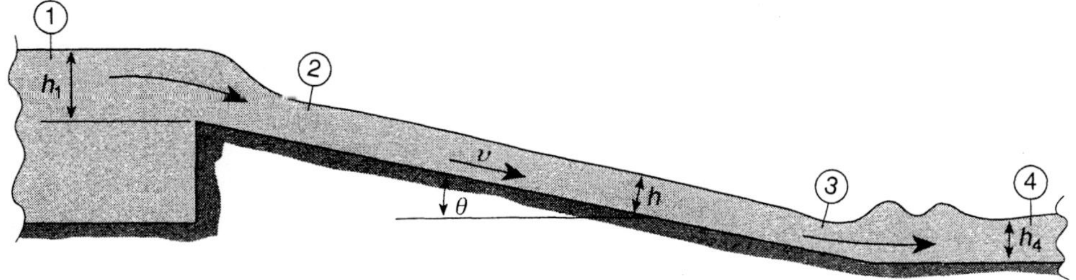

Fig. 4.5 *Flow through a channel*

---

[1] Because the slope is usually very small, there is no significant difference if the run is measured along the incline or along the horizontal, that is, whether $S = \sin \theta$ or $S = \tan \theta$.

If the heights h_1 and h_4 remain constant (or nearly so) the flow will be **steady** and not changing with time. At positions ① and ④ the liquid is at rest (or moving with very low velocity). Between ① and ② the liquid accelerates from rest to some velocity v. If the channel is a relatively long one, velocity v remains constant between positions ② and ③ and the height h of the liquid in the channel also remains constant. Between positions ③ and ④ the liquid decelerates to rest again and there is turbulence as the kinetic energy of the liquid is dissipated.

The acceleration of the fluid, the constant velocity motion, and the deceleration to rest again is analogous to the motion of a ball rolling from a flat surface down an inclined plane to terminal velocity and then back to rest again along another flat surface.

In this book, analysis of channel flow is restricted to situations involving steady flow with constant velocity only (that is for flow between ② and ③). Unsteady flow and channel flow with accelerating motion are not considered (that is for flow between ① and ② or ③ and ④).

4.2 *CHANNEL FLOW USING THE DARCY FORMULA*

Problems involving steady flow in channels can be solved using the Darcy formula for head loss in a similar way to problems solved in Chapter 3 for the free flow of liquids in pipes connected to reservoirs or tanks.

Consider steady flow with constant velocity down a section of channel as shown in Figure 4.6.

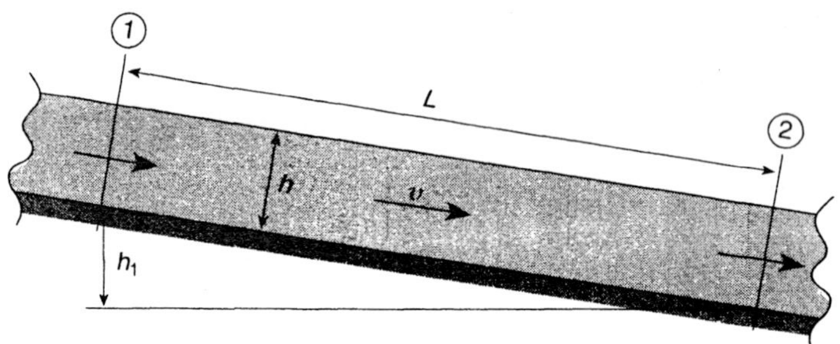

Fig. 4.6 *Section of a channel in steady flow*

Applying the Bernoulli equation between ① and ②:

$$\frac{p_1}{\rho g} + \frac{v_1^{\,2}}{2g} + h_1 + H = \frac{p_2}{\rho g} + \frac{v_2^{\,2}}{2g} + h_2 + H_L$$

Now $p_1 = 0$ and $p_2 = 0$ (atmospheric pressure),
and $v_1 = v_2 = v$ (steady flow with constant velocity).
Also $h_2 = 0$ (datum) and $H = 0$ (there is no pump or turbine between ① and ②). Therefore the Bernoulli equation reduces to:

$$h_1 = H_L$$

That is, the drop in vertical height equals the head loss.

The slope of the channel $S = \dfrac{h_1}{L}$

$$\therefore LS = h_1$$
$$\therefore LS = H_L$$

The Darcy formula for a circular pipe running full is:

$$H_L = f\,\frac{L}{d}\,\frac{v^2}{2g}$$

For a non-circular section, or a circular section not running full, the equivalent diameter d_e is used in place of d. As given in Chapter 1, d_e is defined by Formula 1.2:

$$d_e = \frac{4A}{P}$$

where A = sectional area in flow, and P = wetted perimeter.
For channel flow, using d_e in place of d, the Darcy formula is:

$$H_L = f\,\frac{L}{d_e}\,\frac{v^2}{2g}$$
$$\therefore LS = f\,\frac{L}{d_e}\,\frac{v^2}{2g}$$
$$S = \frac{f}{d_e}\,\frac{v^2}{2g}$$
$$\text{or} \qquad v^2 = \frac{Sd_e\,2g}{f}$$

$$\boxed{\;\therefore v = \sqrt{\frac{Sd_e\,2g}{f}}\;}$$

Darcy formula applied to channel flow (4.1)

Known friction factor

If the friction factor is known, a direct solution is possible using Formula 4.1. A problem of this type is given as Example 4.1.

Example 4.1

Liquid flows down a pipe of diameter 200 mm with slope 1 in 100, as shown in Figure 4.7. The liquid depth in the pipe is 150 mm. If the friction factor is 0.02, determine the steady flow velocity.

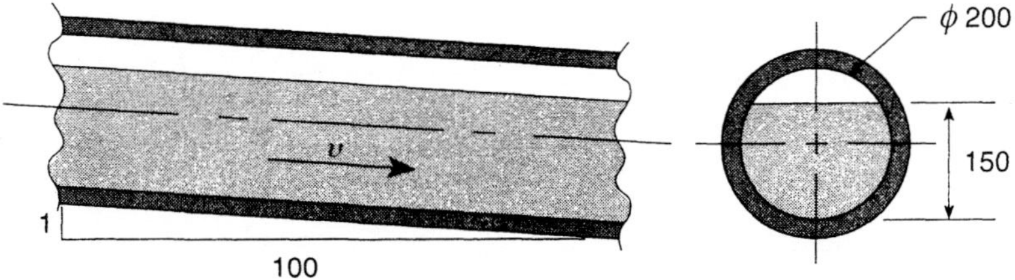

Fig. 4.7

Solution

Before applying Formula 4.1 it is necessary to calculate the equivalent diameter d_e. Refer to Figure 4.8.

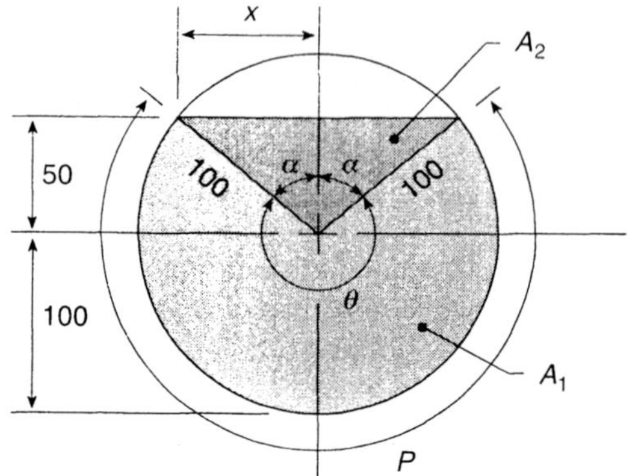

Fig. 4.8

$$\cos \alpha = \frac{50}{100}$$
$$\therefore \alpha = 60° \text{ and } \theta = 240°$$
$$x = 100 \sin \alpha = 86.6 \text{ mm}$$
$$A_1 = \frac{240}{360} \times A(200) = 20\,944 \text{ mm}^2$$
$$A_2 = 50x = 50 \times 86.6 = 4330 \text{ mm}^2$$
$$\therefore A = 20\,944 + 4330 = 25\,274 \text{ mm}^2$$
$$\text{Perimeter } P = \frac{240}{360} \times \pi \times 200 = 418.88 \text{ mm}$$
$$d_e = \frac{4A}{P} = \frac{4 \times 25\,274}{418.88} = 241.35 \text{ mm}$$

Applying Formula 4.1:

$$v = \sqrt{\frac{S d_e 2g}{f}}$$

with $S = 0.01$ (1/100), $d_e = 0.24135$ m, $f = 0.02$

$$\therefore v = \sqrt{\frac{0.01 \times 0.24135 \times 19.62}{0.02}}$$
$$\therefore v = \mathbf{1.54 \text{ m/s}}$$

Unknown friction factor

If the friction factor is not known (or given) a direct solution using Formula 4.1 is not possible and an iterative (trial-and-error) approach is needed. In most cases flow is turbulent, and hence the Moody formula may be used to calculate the friction factor. Alternatively the Moody diagram may be used.

When a trial-and-error approach is used it simplifies the calculations if the factors that do not vary are inserted in the relevant equations and the equations simplified before trial values are inserted. This approach is adopted in Example 4.2 below as this is an example of a channel flow problem with an unknown friction factor.

Example 4.2

The liquid of Example 4.1 has a viscosity of 2×10^{-3} Pas and a relative density of 0.8. The pipe is made of steel. Determine the steady flow velocity.

Solution

From Appendix 4, ε for commercial steel = 0.045 mm
From Example 4.1, d_e = 241.35 mm

$$\varepsilon_R = \frac{\varepsilon}{d_e} = \frac{0.045}{241.35} = 0.1865 \times 10^{-3}$$

$$Re = \frac{v d_e \rho}{\mu} = \frac{v \times 0.24135 \times 800}{2 \times 10^{-3}} = 0.0965 v \times 10^6$$

Trial 1

It is logical to use, as the initial trial value, the velocity 1.54 m/s determined in Example 4.1:

$$Re = 0.0965 \times 1.54 \times 10^6 = 0.149 \times 10^6 \text{ (and flow is turbulent)}$$

$$f = 0.0055 \left[1 + \left(20\,000 \varepsilon_R + \frac{10^6}{Re} \right)^{\frac{1}{3}} \right]$$

$$= 0.0055 \times \left[1 + \left(20 \times 0.1865 + \frac{1}{0.149} \right)^{\frac{1}{3}} \right]$$

$$= 0.0055 \times \left[1 + (3.729 + 6.726)^{\frac{1}{3}} \right]$$

$$= 0.0175$$

Applying Formula 4.1:

$$v = \sqrt{\frac{S d_e 2g}{f}}$$

$$v = \sqrt{\frac{0.01 \times 0.24135 \times 19.62}{f}}$$

$$v = \sqrt{\frac{0.0474}{f}}$$

Substituting $f = 0.0175$ gives $v = 1.64$ m/s

Trial 2

$$v = 1.64 \text{ m/s}$$

$$Re = 0.0965 \times 1.64 \times 10^6 = 0.1586 \times 10^6$$

$$f = 0.0055 \times \left[1 + \left(3.729 + \frac{1}{0.1586} \right)^{\frac{1}{3}} \right]$$

$$= 0.0174$$

Applying Formula 4.1:

$$v = \sqrt{\frac{0.0474}{0.0174}}$$

$$= 1.65 \text{ m/s}$$

Trial 3

$$v = 1.65 \text{ m/s}$$
$$Re = 0.0965 \times 1.65 \times 10^6 = 0.1594 \times 10^6$$

$$f = 0.0055 \times \left[1 + \left(3.729 + \frac{1}{0.1594} \right)^{\frac{1}{3}} \right]$$

$$= 0.0173$$

Applying Formula 4.1:

$$v = \sqrt{\frac{0.0474}{0.01735}}$$

$$= 1.65 \text{ m/s}$$

Therefore the solution (to three significant figures) is v = **1.65 m/s**.

Self-test problem 4.1

The slope of the pipe in Example 4.2 is now to be changed so that the liquid velocity will be at least 1 m/s when the pipe is running half-full. Determine the minimum slope required.

4.3 THE CHEZY FORMULA

One of the oldest special formulas for channel flow was proposed by the French researcher Chezy in 1775. The formula, which he established by experiment, is:
where:

$$\boxed{v = C\sqrt{RS}}$$

Chezy formula (4.2)

v = steady flow velocity (m/s)
R = hydraulic radius (m)
S = slope of the channel (no units)
C = Chezy coefficient ($m^{\frac{1}{2}}\ s^{-1}$).

The Chezy coefficient is similar to the friction factor and is a coefficient that takes into account the frictional resistance of the channel surfaces. However, while friction factor has no units, the Chezy coefficient has units as stated. The hydraulic radius R is defined as:

$$R = \frac{A}{P}$$

where A = sectional area in flow and P = wetted perimeter.

Since the equivalent diameter is defined as:

$$d_e = \frac{4A}{P}$$

it is clear that:

$$R = \frac{A}{P} = \frac{d_e}{4}$$

Hydraulic radius (4.3)

Note   Do not confuse hydraulic radius with the radius of a circle because for a circular section the radius is half the diameter, whereas the hydraulic radius is one-quarter the diameter.

Comparison of the Chezy and Darcy formulas for channel flow

The Chezy formula for channel flow is:

$$v = C\sqrt{RS}$$

The Darcy formula applied to channel flow is:

$$v = \sqrt{\frac{Sd_e 2g}{f}}$$

$$\therefore C\sqrt{RS} = \sqrt{\frac{Sd_e 2g}{f}}$$

$$\therefore C^2 RS = \frac{Sd_e 2g}{f}$$

S cancels out and $R = \dfrac{d_e}{4}$

$$\therefore C^2 \frac{d_e}{4} = \frac{d_e 2g}{f}$$

Now d_e cancels out, leaving:

$$\frac{C^2}{4} = \frac{2g}{f}$$

$$\therefore C = \sqrt{\frac{8g}{f}}$$

Chezy coefficient (4.4)

Thus the Chezy formula is in fact a special application of the more general Darcy formula for head loss in a pipe (even though it preceded it historically by more than a century).[2]

[2] This is similar to the way in which the Torricelli equation for free flow from a tank through a nozzle or orifice (discovered experimentally) is in fact a special application of the more general Bernoulli equation which was formulated many years later.

Example 4.3

Solve Example 4.1 using the Chezy formula.

Solution

Using Formula 4.4, with $f = 0.02$:

$$C = \sqrt{\frac{8g}{f}} = \sqrt{\frac{8 \times 9.81}{0.02}} = 62.64 \ \text{m}^{\frac{1}{2}} \ \text{s}^{-1}$$

From Example 4.1, $d_e = 241.35$ mm

Now $R = \dfrac{d_e}{4} = 60.38$ mm $= 0.06038$ m

Using Formula 4.3 with $S = 0.01$:

$$\begin{aligned}
v &= C\sqrt{RS} \\
&= 62.64 \times \sqrt{0.06038 \times 0.01} \\
&= \textbf{1.54 m/s} \ \text{(which is the same answer as obtained previously)}
\end{aligned}$$

Unknown liquid depth

When the depth of liquid in the channel is not known, the solution is more complex and usually produces a polynomial equation, best solved by trial and error. A typical example is given in Example 4.4.

Example 4.4

An open rectangular channel 4 m wide has a slope of 1 in 800 and conveys liquid at a rate of 8 m³/s. The Chezy coefficient may be taken to have a constant value of 50 m$^{\frac{1}{2}}$ s⁻¹.

(a) Determine the depth of liquid in the channel, accurate to three significant figures.
(b) Determine the velocity of the liquid in the channel from the flow rate and depth obtained in (a).
(c) Check the velocity, using the Chezy formula.

Solution

(a) Referring to Figure 4.9, $R = \dfrac{A}{P} = \dfrac{4h}{4 + 2h}$

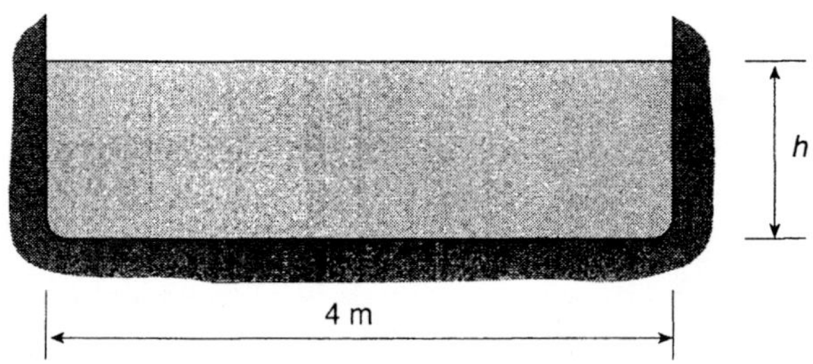

Fig. 4.9

Now

$$\begin{aligned}
\dot{V} &= vA \\
\therefore 8 &= v \times 4h \\
\therefore v &= \frac{2}{h}
\end{aligned}$$

The Chezy formula is:

$$v = C\sqrt{RS}$$

$$\therefore \frac{2}{h} = 50\sqrt{\frac{4h}{4 + 2h}} \times \frac{1}{800}$$

Squaring both sides:

$$\frac{4}{h^2} = 2500 \times \frac{4h}{4 + 2h} \times \frac{1}{800}$$

$$\therefore \frac{1.28}{h^2} = \frac{4h}{4 + 2h}$$

$$\therefore 5.12 + 2.56h = 4h^3$$

$$\therefore h^3 - 0.64h - 1.28 = 0$$

This equation may be solved by trial and error

Trial 1
$$h = 1 \qquad LHS = -0.92$$

Trial 2
$$h = 1.1 \qquad LHS = -0.653$$

Trial 3
$$h = 1.2 \qquad LHS = -0.32$$

Trial 4
$$h = 1.3 \qquad LHS = +0.085$$
Therefore the solution is that h lies between 1.2 and 1.3 m, and is closer to 1.3 m

Trial 5
$$h = 1.29 \qquad LHS = +0.041$$

Trial 6
$$h = 1.28 \qquad LHS = -0.002$$
Therefore the solution, accurate to three significant figures, is h = **1.28 m.**

(b) $v = \dfrac{2}{h} = \dfrac{2}{1.28}$ = **1.56 m/s**

(c) $R = \dfrac{A}{P} = \dfrac{4h}{4 + 2h} = \dfrac{4 \times 1.28}{4 + 2 \times 1.28}$ = 0.7805 m

$$v = C\sqrt{RS} = 50 \times \sqrt{0.7805 \times \frac{1}{800}} = \textbf{1.56 m/s} \text{ (which checks).}$$

Self-test problem 4.2

A vee-shaped channel as illustrated in Figure 4.10 has a slope of 1 in 900 and conveys

liquid at a rate of 5 m³/s. The Chezy coefficient is 60 m$^{\frac{1}{2}}$ s^{-1}. Determine:
(a) the depth of liquid in the channel h (accurate to three significant figures);
(b) the velocity of the liquid in the channel from the flow rate and depth obtained in (a);
(c) the equivalent friction factor.
Also:
(d) check the velocity using the Chezy formula.

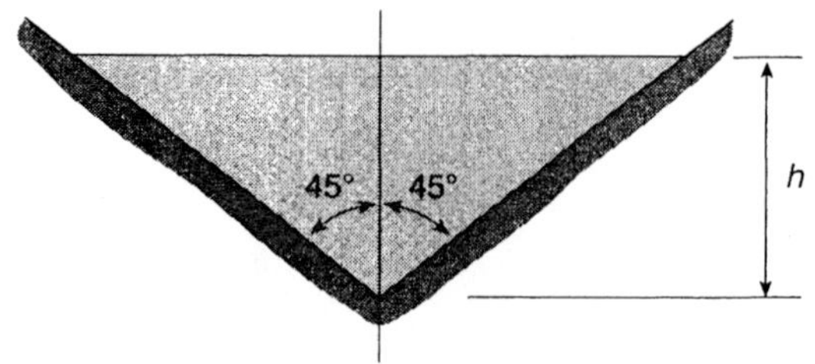

Fig. 4.10

4.4 *LARGE-SECTION CHANNELS*

The difficulty in using the Darcy or Chezy formulas for flow in channels is that the friction factor and Chezy coefficient depend on the velocity. In turn the velocity depends on the friction factor or Chezy coefficient. This interdependence means that a trial-and-error solution is required unless the friction factor or Chezy coefficient is assumed to have a constant value; that is, a value that does not change with velocity.

However, many channels are of large section and are made of relatively rough materials. For these channels, with liquid flowing with typical velocities, flow is turbulent and the friction factor and Chezy coefficient do not vary greatly with the velocity. This is demonstrated in Example 4.5.

Example 4.5

A rectangular channel 1 m wide carries water to a depth of 1 m. The average height of the surface protuberances (absolute roughness) of the channel surfaces is 0.9 mm. Determine:
(a) the Chezy coefficient when the flow velocity is 1 m/s;
(b) the Chezy coefficient when the flow velocity is 2 m/s;
(c) the percentage change in Chezy coefficient between the two velocity values given.
Take the viscosity of the water to be 1×10^{-6} m²/s.

Solution

$$R = \frac{A}{P} = \frac{1 \times 1}{3} = 0.333 \text{ m}$$

$$d_e = \frac{4A}{P} = 1.333 \text{ m}$$

$$\varepsilon_R = \frac{\varepsilon}{d_e} = \frac{0.9}{1333} = 0.675 \times 10^{-3}$$

(a) $v = 1$ m/s

$$Re = \frac{v d_e}{v} = \frac{1 \times 1.3333}{1 \times 10^{-6}} = 1.333 \times 10^6 \qquad \text{(and flow is turbulent)}$$

$$f = 0.0055 \left[1 + \left(20\,000\varepsilon_R + \frac{10^6}{Re} \right)^{\frac{1}{3}} \right]$$

$$= 0.0055 \times \left[1 + \left(20 \times 0.675 + \frac{1}{1.333} \right)^{\frac{1}{3}} \right]$$

$$= 0.0055 \times \left[1 + (13.5 + 0.75)^{\frac{1}{3}} \right]$$

$$= 0.0188$$

$$C = \sqrt{\frac{8g}{f}} = \sqrt{\frac{8 \times 9.81}{0.0188}} = \textbf{64.55 m}^{\frac{1}{2}} \textbf{ s}^{-1}$$

(b) $v = 2$ m/s

$$Re = \frac{vd_c}{\nu} = \frac{2 \times 1.3333}{1 \times 10^{-6}} = 2.667 \times 10^6 \qquad \text{(and flow is turbulent)}$$

$$f = 0.0055 \left[1 + \left(20\,000\varepsilon_R + \frac{10^6}{Re} \right)^{\frac{1}{3}} \right]$$

$$= 0.0055 \times \left[1 + \left(13.5 + \frac{1}{2.667} \right)^{\frac{1}{3}} \right]$$

$$= 0.0055 \times \left[1 + (13.5 + 0.375)^{\frac{1}{3}} \right]$$

$$= 0.0187$$

$$C = \sqrt{\frac{8g}{f}} = \sqrt{\frac{8 \times 9.81}{0.0187}} = \mathbf{64.75\ m^{\frac{1}{2}}\ s^{-1}}$$

(c) The percentage change $= \dfrac{64.75 - 64.55}{64.55} \times 100 = \mathbf{0.31\%}$

4.5 *THE MANNING FORMULA*

The calculation given in Example 4.5 shows that if the channel is relatively rough, and the Reynolds number high, the Chezy coefficient is almost independent of the velocity. In this example the velocity was doubled but the Chezy coefficient changed by less than $\frac{1}{3}\%$. This change is negligible with regard to the order of accuracy usually obtained in channel flow calculations. This conclusion was arrived at in 1890 from analysis of experimental data by Robert Manning. He found that for large channels and large open pipes the Chezy coefficient was almost independent of the velocity, but was dependent on the roughness of the channel and the hydraulic radius. He obtained the following empirical relationship:

$$\boxed{C = \frac{R^{\frac{1}{6}}}{n}} \qquad\qquad \text{Chezy coefficient (4.5)}$$

where n is a roughness coefficient with units $m^{-\frac{1}{3}}$ s. The greater the roughness of the channel, the higher is the value of n.

The Chezy formula is:

$$v = C\sqrt{RS}$$

Substituting the Manning value for C:

$$v = \frac{R^{\frac{1}{6}}R^{\frac{1}{2}}S^{\frac{1}{2}}}{n}$$

$$\therefore \boxed{v = \frac{1}{n}\,R^{\frac{2}{3}}S^{\frac{1}{2}}} \qquad\qquad \text{Manning formula (4.6)}$$

This is known as the Manning formula.

Values of the Manning coefficient n have been determined experimentally. Typical average values for *water* flowing through channels made of rough materials are listed in Appendix 9.

Relationship between the Manning coefficient and the friction factor

Formula 4.5 gives the relationship between the Manning coefficient and the Chezy coefficient. As has been seen, the relationship between the Chezy coefficient and the friction factor is given by Formula 4.4. Therefore the relationship between the Manning coefficient and the friction factor can be derived by combining these two formulas.

$$C = \frac{R^{\frac{1}{6}}}{n} \quad \text{(Formula 4.5)}$$

$$C = \sqrt{\frac{8g}{f}} \quad \text{(Formula 4.4)}$$

$$\therefore \quad \frac{R^{\frac{1}{6}}}{n} = \sqrt{\frac{8g}{f}}$$

$$\therefore \quad \boxed{n = R^{\frac{1}{6}} \sqrt{\frac{f}{8g}}} \qquad \text{Manning coefficient (4.7)}$$

Note Formula 4.7 may be used to determine the Manning coefficient from the friction factor. However, tabulated values of the Manning coefficient are usually obtained from experiments with water flow in channels.

Example 4.6

Water flows through a masonry channel with trapezoidal section as illustrated in Figure 4.11. The depth of water in the channel is 800 mm and the slope of the channel is 1 in 2000. Determine:
(a) the discharge (using the Manning formula);
(b) the equivalent Chezy coefficient;
(c) the equivalent friction factor.

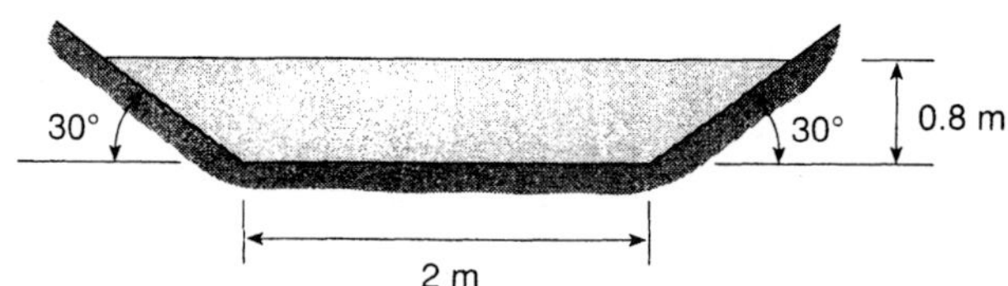

Fig. 4.11

Solution

(a) First determine some distances by trigonometry, as shown in Figure 4.12 opposite:

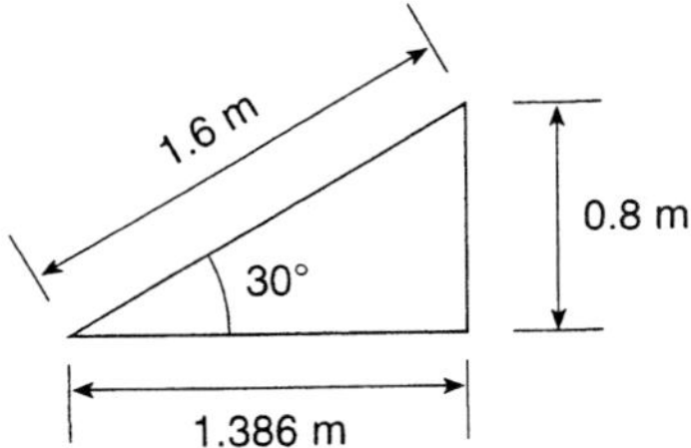

Fig. 4.12

$$A = 2 \times 0.8 + 0.8 \times 1.386 = 2.7085 \text{ m}^2$$
$$P = 2 + 2 \times 1.6 = 5.2 \text{ m}$$
$$R = \frac{A}{P} = \frac{2.7085}{5.2} = 0.521 \text{ m}$$

From Appendix 9, for a masonry channel $n = 0.018 \text{ m}^{-\frac{1}{3}} \text{ s}$
Using the Manning formula:

$$v = \frac{1}{n} R^{\frac{2}{3}} S^{\frac{1}{2}}$$
$$= \frac{1}{0.018} \times 0.521^{\frac{2}{3}} \times \frac{1}{2000^{\frac{1}{2}}}$$
$$= 0.804 \text{ m/s}$$
$$\dot{V} = vA = 0.804 \times 2.7085 = \mathbf{2.18 \text{ m}^3/s}$$

(b) Applying Formula 4.5:

$$C = \frac{R^{\frac{1}{6}}}{n} = \frac{0.521^{\frac{1}{6}}}{0.018} = \mathbf{49.8 \text{ m}^{\frac{1}{2}} \text{ s}^{-1}}$$

(c) Applying Formula 4.7:

$$n = R^{\frac{1}{6}} \sqrt{\frac{f}{8g}}$$
$$\therefore 0.018 = 0.521^{\frac{1}{6}} \frac{\sqrt{f}}{\sqrt{8 \times 9.81}}$$
$$\therefore f = \mathbf{0.0316}$$

Self-test problem 4.3

A rectangular channel carrying water is made of smooth cement and is 3 m wide.
Determine:
(a) the necessary slope of the channel in order that the discharge is 5 m^3/s when the depth
is 1 m;
(b) the equivalent Chezy coefficient.

4.6 OPTIMUM SECTION SHAPE

The optimum section shape for any duct or channel conveying a fluid is the *shape that
gives maximum cross-sectional area for minimum perimeter in contact with the fluid.*
That is, the optimum section shape is the one where the ratio *A/P* is largest. The ratio

A/P is the hydraulic radius R, therefore the optimum section shape for the transport of a fluid is the shape that has the largest hydraulic radius R. This is true whether the fluid is flowing under pressure (pipe flow) or flowing with a free surface (channel flow). There are several reasons for this:

- the smallest quantity of material is necessary in the construction of the duct or channel and therefore material costs are minimised;
- the frictional losses are reduced and less slope is required because fluid friction reduces when the surface area in contact with the fluid is reduced;
- the cross-sectional area is minimised and this gives the best utilisation of available space.

In the case of a fluid under pressure, the optimum section shape is circular. This is because for a given cross-sectional area, the circle has the greatest hydraulic radius of all possible sections.

In the case of a fluid not under pressure (channel flow), a semicircle is the optimum section shape[3]. For ease of construction, channel sections are often rectangular or trapezoidal. For these channels the optimum section occurs when the shape of the channel most closely approximates a semicircle.

Rectangular channel optimum section occurs when the depth is one-half the breadth[3]. *Trapezoidal channel* optimum section occurs when the shape is one-half hexagon[3]. The various optimum section shapes for channels are illustrated in Figure 4.13.

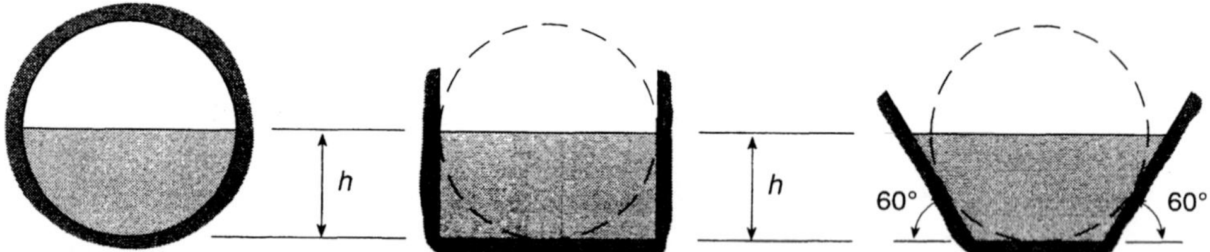

Fig. 4.13   *Optimum section shapes for various channels*

Often the rate of flow through a channel varies. When this is the case the trapezoidal section generally provides the best compromise between the smaller and the higher flow rates. In some cases there can be a very large difference between maximum and minimum flow rates. For example, when a channel conveys stormwater it will have a very high flow rate during times of heavy rain. This flow rate occurs infrequently. Usually the flow rate will be small, and during dry periods the flow rate could amount to only a trickle. In such circumstances a compound channel shape, as illustrated in Figure 4.14, provides the optimum section.

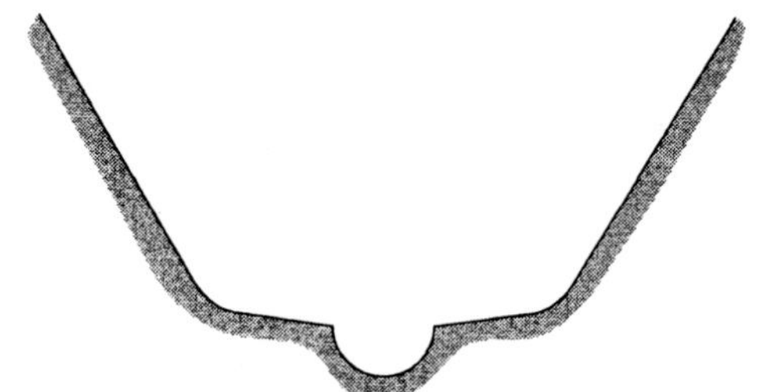

Fig. 4.14   *Optimum section shape with very large range of flow rates*

---

[3] The proof of these statements requires the application of calculus and is not given in this book. However, Problems 4.3 and 4.4 at the end of this chapter illustrate the proof of this for circular and rectangular sections.

Note In practice, channel shapes are often dictated by the materials and methods of construction and are not necessarily optimum. For example, for a natural earth or gravel channel the only practical section shape is trapezoidal, with sides inclined to the horizontal at considerably less than 60°.

Summary

Channel flow occurs when a liquid flows down an inclined pipe or duct with a free upper surface. Channel flow is different to pipe or duct flow where fluid flows under pressure, because in channel flow the upper surface of the fluid is not in contact with the wall of the pipe or duct. Hence in channel flow the fluid flows because of gravity forces only and is not driven by higher upstream pressure forces compared to the downstream ones. Therefore channel flow occurs only when the pipe or duct is inclined to the horizontal, and flow always takes place in the downward direction. Hence channel flow cannot be used when it is necessary to raise a liquid, for example when water is pumped up from a well or reservoir. However, the great advantage of channel flow as a method of moving fluids is that no pump is needed, and no power input is necessary.

Channel flow problems can be solved in the same way as any other fluid flow problem in a duct or pipe by application of the Bernoulli equation and Darcy formula. In channel flow the Bernoulli equation reduces to: $h_1 = H_L$, that is, the change in elevation between points ① and ② is equal to the head loss. By applying the Darcy formula for head loss, Formula 4.1 is obtained:

$$v = \sqrt{\frac{S d_e \, 2g}{f}}$$

Because the Darcy formula for head loss applies to a circular pipe with diameter d running full, for channel flow problems the equivalent diameter d_e as defined by Formula 4.2 must be used in place of d:

$$d_e = \frac{4A}{P}$$

If the friction factor is given, calculation of the flow velocity using Formula 4.1 is straightforward. If the friction factor is not known, but the surface roughness is known, the Moody formula may be used to calculate the friction factor, because flow is almost always turbulent. However, an iterative approach is necessary because the Reynolds number and friction factor depend on the velocity.

One of the earliest special formulas developed for channel flow is the Chezy formula, Formula 4.2:

$$v = C\sqrt{RS}$$

One of the terms in this formula is the hydraulic radius R, defined as the cross-sectional area of the fluid, divided by the fluid perimeter in contact with the channel. The hydraulic radius is also equal to the equivalent diameter divided by four. Note that the larger the cross-section of the channel, the larger is the value of R and the higher is the velocity (for a given slope). Put another way, less slope is needed in a large channel to obtain a certain velocity than in a small channel.

The Chezy formula enables direct calculation of the flow velocity in the channel if the Chezy coefficient is known. However, the Chezy coefficient is related to the friction factor, which in turn varies with the velocity, so direct solution is not possible if the Chezy coefficient is not given. The Chezy coefficient is related to the friction factor by Formula 4.4:

$$C = \sqrt{\frac{8g}{f}}$$

Many channel flow problems involve large-section channels made of relatively rough materials. For channels of this type the Chezy coefficient varies only slightly with the velocity of fluid in the channel. This is shown by example in the text, but was discovered experimentally by Robert Manning. He derived a relationship for the Chezy coefficient based only on the roughness of the channel surface and the hydraulic radius. This relationship is given by Formula 4.5:

$$C = \frac{R^{\frac{1}{6}}}{n}$$

In this formula the roughness of the channel is expressed in terms of a coefficient n. Values of n have been determined experimentally for the flow of water through channels constructed of different types of materials. Average values are given in Appendix 9. These values apply to water only and should not be used for other types of liquid.

Formula 4.5 can be substituted in the Chezy formula to derive the Manning formula (Formula 4.6):

$$v = \frac{1}{n} R^{\frac{2}{3}} S^{\frac{1}{2}}$$

Just as there is a relationship between the Chezy coefficient and the friction factor, so there is a relationship between the Manning coefficient and the friction factor. It is given by Formula 4.7:

$$n = R^{\frac{1}{6}} \sqrt{\frac{f}{8g}}$$

As there are three formulas that can be used to determine fluid velocity in a channel; the Darcy, Chezy and Manning formulas; the question arises as to which one is most appropriate in each situation. The following summary is a guide:

- Use the Darcy formula if the friction factor or absolute roughness (height of the roughness protuberances) is given. The Darcy formula is valid for any shape or size of channel and therefore may be used for relatively small-section channels.
- Use the Chezy formula if the Chezy coefficient is given and is assumed to have a constant value. This is a good assumption for large-section channels, particularly if the surface is relatively rough.
- Use the Manning formula if the Manning coefficient is given and is assumed to have a constant value. This is a good assumption for large-section channels,

particularly if the surface is relatively rough. If the Manning coefficient is not given, but the nature of the surfaces of the channel is specified, the Manning coefficient can be obtained from tabulated values for the coefficient such as those listed in Appendix 9. These values apply only to channels where water flows.

When the channel conveys fluid at a fixed rate of flow, the optimum channel section shape is that of a semicircle. For any other channel shape the optimum section occurs when the channel shape most closely resembles a semicircle. Hence for a rectangular channel the optimum section shape is when the depth is one-half the breadth, and for a trapezoidal section, when the shape is one-half hexagon. Note that for a semicircular shape the equivalent diameter is the same as the diameter, and the hydraulic radius is therefore the diameter divided by four.

When a channel conveys a fluid with varying rates of flow, a trapezoidal section is usually optimum. If there is a very large range of flow rates, a compound section such as a trapezoidal section with a semicircular lower portion provides a good compromise. Channel sections are often dictated by the materials and construction methods available, and optimum sections are not always practical.

 # Problems

Note For these problems, assume steady flow with constant velocity and height of liquid in the channel. Use Appendix 4 for roughness values and Appendix 9 for Manning coefficients when these are required.

4.1 (a) Explain the two reasons why rain falling on a sloped galvanised iron roof and running down it is *not* an example of steady channel flow with constant velocity.

   (b) What is meant by the optimum section shape of a channel, and under what conditions does it occur?

   (c) What is the optimum section shape of a rectangular channel?

   (d) What is the optimum section shape of a trapezoidal channel?

   (e) For a sloped metal roof, which of the following shapes will give the highest flow rate: flat, box section (cliplock), corrugated iron? Explain your answer.

4.2 (a) State in words the result of applying the Bernoulli equation to steady flow in a channel.

   (b) What assumption about the flow is inherent in the Chezy and Manning formulas?

   (c) Under what conditions is this assumption valid?

   (d) What is the optimum depth of liquid in a circular channel not running full?

4.3 Determine the hydraulic radius for a rectangular section channel with cross-sectional area 1 m^2 when the water depth is:

   (a) 1.5 times the width;

   (b) equal to the width;

   (c) 0.5 times the width;

   (d) 0.25 times the width.

   Also:

   (e) draw conclusions from the above calculations.

   (a) 0.3062 m (b) 0.3333 m (c) 0.3536 m (d) 0.3333 m

4.4 Determine the radius and hydraulic radius for a circular-shaped channel with water cross-sectional area of 1 m^2 when the water depth is:

   (a) 1.5 times the radius;

   (b) equal to the radius;

   (c) 0.5 times the radius.

Also:

(d) draw conclusions from the above calculations.

    (a) r = 0.629 m, R = 0.3795 m (b) r = 0.7979 m, R = 0.399 m

    (c) r = 1.276 m, R = 0.3742 m

4.5 A pipe of diameter 300 mm with a slope of 1 in 200 conveys liquid. The friction factor is 0.02. Determine the velocity and flow rate when the liquid depth in the pipe is:

(a) 150 mm;

(b) 180 mm.

    (a) 1.21 m/s, 42.9 L/s (b) 1.28 m/s, 56.6 L/s

4.6 A cast-iron pipe with slope 1 in 200 and diameter 600 mm conveys water. If the pipe is half-full, determine the velocity and discharge using the Manning formula.

    1.33 m/s, 188 L/s

4.7 A rectangular channel carries 2 m^3/s of water with velocity 1.2 m/s. The depth is one-half the width and the Manning coefficient is 0.02. Determine:

(a) the width of the channel;

(b) the depth of water in the channel;

(c) the slope of the channel.

    (a) 1.83 m (b) 0.913 m (c) 1/610

4.8 Liquid with viscosity 20×10^{-6} m^2/s flows through a cast-iron pipe of diameter 500 mm which drops 200 mm over a length of 100 m. Determine the velocity, friction factor, and discharge (accurate to three significant figures) when the pipe is half-full. Use the Darcy formula with an initial trial value of 1 m/s for the velocity.

    Trial 1: v = 1 m/s, f = 0.0258

    Trial 2: v = 0.873 m/s, f = 0.0265

    Trial 3: v = 0.860 m/s, f = 0.0266

    Trial 4: v = 0.859 m/s, f = 0.0266

    Solution: v = 0.859 m/s, f = 0.0266, $\dot{V}$ = 84.3 L/s

4.9 **(a)** In Problem 4.8, what is the equivalent Chezy coefficient? Use the Chezy formula to determine the velocity.

    (b) In Problem 4.8, what is the equivalent Manning coefficient? Use the Manning formula to determine the velocity.

    (a) 54.3 m$^{\frac{1}{2}}$ s^{-1}, 0.859 m/s (b) 0.013 m$^{-\frac{1}{3}}$ s, 0.859 m/s

4.10 A channel is vee shaped with each side inclined at 45° to the vertical. The depth of water in the channel is 250 mm and the flow rate is 50 L/s. The Chezy coefficient is 50 m$^{\frac{1}{2}}$ s^{-1}. Determine:

(a) the velocity of the water;

(b) the slope of the channel;

(c) the equivalent friction factor.

    (a) 0.8 m/s (b) 1/345 (c) 0.0314

4.11 In Problem 4.10, what is the equivalent Manning coefficient? Determine the channel slope using the Manning formula.

    0.0133 m$^{-\frac{1}{3}}$ s, 1/345

4.12 A rectangular channel has a width of 4 m and a slope of 1 in 500. Water with viscosity 1.15×10^{-6} m^2/s flows through the channel. The absolute roughness of the channel surfaces is 1.2 mm. Determine the velocity, friction factor, and discharge (accurate to three significant figures) when the depth in the channel is 1.2 m. Use the Darcy formula with an initial trial value of 1 m/s for the velocity.

Trial 1: $v = 1$ m/s, $f = 0.0167$
Trial 2: $v = 2.66$ m/s, $f = 0.0166$
Trial 3: $v = 2.67$ m/s, $f = 0.0166$
Solution: $v = 2.67$ m/s, $f = 0.0166$, $\dot{V} = 12.8$ m³/s

4.13 For Problem 4.12, determine the equivalent:
   (a) Chezy coefficient;
   (b) Manning coefficient.
   Also:
   (c) determine the discharge using the Chezy formula.

   (a) 68.8 m$^{\frac{1}{2}}$ s^{-1} (b) 0.0139 m$^{-\frac{1}{3}}$ s (c) 12.8 m³/s

4.14 A natural-earth canal is trapezoidal and has sides which slope at an angle of 25° to the horizontal. The width of the bottom of the canal is 2.5 m. If the slope is 1 in 5000, determine the flow rate of water through the canal when the depth of water in the canal is 1 m.

   1.96 m³/s

4.15 For the canal in Problem 4.14, if it is desired that the canal discharge 3 m³/s of water when the depth is the same, what is the required slope of the canal?

   1/2125

4.16 An irrigation canal is trapezoidal and 2 m wide at the bottom. The sides are inclined at 60° to the horizontal and the slope of the canal is 1 in 5000. It carries water with velocity 0.5 m/s. The Manning coefficient is 0.02 m$^{-\frac{1}{3}}$ s. Determine:
   (a) the hydraulic radius;
   (b) the depth of water in the canal;
   (c) the flow rate.
   (a) 0.5946 m (b) 0.992 m (c) 1.275 m³/s

4.17 An open rectangular channel 3 m wide has a slope of 1 in 900 and conveys liquid at a rate of 5 m³/s. The Chezy coefficient may be taken to have a constant value of 60 m$^{\frac{1}{2}}$ s^{-1}.
   (a) Determine the depth of liquid in the channel, accurate to three significant figures.
   (b) Determine the velocity of the liquid in the channel from the flow rate and the depth obtained in (a).
   (c) Check the velocity using the Chezy formula.
   (d) Determine the equivalent friction factor.
   (e) Determine the equivalent Manning coefficient.
   (a) 1.06 m (b) 1.57 m/s (c) 1.58 m/s (d) 0.0218 (e) 0.0154 m$^{-\frac{1}{3}}$ s

4.18 A rectangular masonry canal 6 m wide carrying water with viscosity 1.1×10^{-6} m²/s has a slope of 1 in 10 000. The flow rate is 12 m³/s. Determine:
   (a) the depth of water in the canal (accurate to three significant figures);
   (b) the velocity in the canal.
   (a) 2.81 m (b) 0.718 m/s

4.19 For the canal in Problem 4.18, determine the equivalent:
   (a) Chezy coefficient;
   (b) friction factor;
   (c) absolute roughness.
   (a) 59.1 m$^{\frac{1}{2}}$ s^{-1} (b) 0.0225 (c) 8.49 mm

4.20 A rectangular channel with width 2 m has a surface roughness of 0.2 mm and a slope of 1/1000. It carries water with viscosity 1.15×10^{-6} m²/s. Determine the friction factor and the velocity of the water in the channel, accurate to 3 significant figures, when the water depth is 2 m. Use the Darcy formula with an initial trial value for the friction factor of 0.02.

Trial 1: $f = 0.02$, $v = 1.62$ m/s
Trial 2: $f = 0.0121$, $v = 2.075$ m/s
Trial 3: $f = 0.0121$, $v = 2.08$ m/s
Solution: $f = 0.0121$, $v = 2.08$ m/s

4.21 For Problem 4.20, determine the equivalent:
(a) Chezy coefficient;
(b) Manning coefficient.

(a) 80.5 m$^{\frac{1}{2}}$ s^{-1} (b) 0.0116 m$^{-\frac{1}{3}}$ s

Fluid machinery

Objectives

On completion of this chapter you should be able to:

- describe the various types of fluid machinery and their method of operation;
- outline the advantages/disadvantages of positive-displacement machines compared to rotodynamic machines;
- calculate specific speed, and use it to indicate the most suitable type of pump for a given performance;
- describe cavitation in a pump and the influence that net positive suction head has on it; also calculate the net positive suction head, given the inlet conditions to a pump;
- read and interpret performance curves for positive-displacement and rotodynamic pumps;
- calculate changes in performance when the speed of a pump changes, and redraw the performance curves.

Introduction

So far in this book fluid flow has been analysed in systems where the flow was *free* or *natural flow*, that is flow that occurred naturally. For example, in Chapter 4, channel flow was analysed. In channel flow the fluid is a liquid which flows from a higher to a lower elevation through an open inclined channel. Some examples of free flow are illustrated in Figure 5.1 on page 106.

In Case (a) the fluid is a liquid and it is flowing down the channel because the upstream elevation h_1 is greater than the downstream elevation h_2.

In Case (b) the fluid is a liquid and it is flowing through the pipe connecting two tanks because the surface level h_1 in one tank is greater than the surface level h_2 in the other.

In Case (c) the fluid is a gas and it is flowing out of the container because the pressure in the container p_1 is greater than the outside pressure p_2.

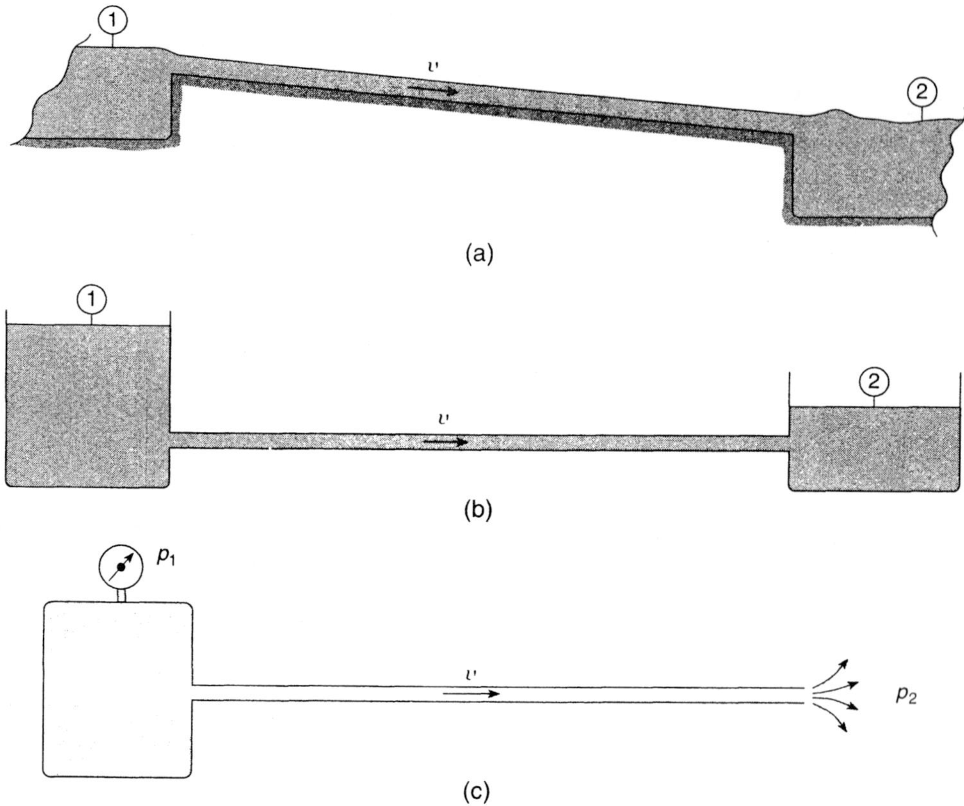

Fig. 5.1   *Free fluid flow*

It can be seen that *free flow of a fluid always occurs from the higher elevation or pressure to a lower elevation or pressure.* That is, a river flows downhill and not uphill, and air flows outward and not inward from a punctured tyre.

When free flow of a fluid occurs there is a waste of energy because fluid energy is converted into internal energy by friction and turbulence. This causes an increase in the fluid temperature, but the heat energy is not usable because the temperature difference is small. However, fluid energy can be recovered (converted to useful mechanical energy) by the use of fluid machinery as shown in Figure 5.2 (opposite). In Case (a) a turbine is installed in a pipe in which liquid is flowing from a higher to a lower elevation. In Case (b) a motor is installed in a pipe in which gas is flowing from a higher to a lower pressure. Indeed, the conversion of fluid power to mechanical power was the first means by which useful amounts of mechanical power were generated without using human or animal muscular energy. The earliest fluid machines for this purpose were waterwheels and windmills. These furnished useful amounts of power but were inefficient, and have been replaced in modern times by efficient water turbines and wind-driven propellers.

In many applications in fluid mechanics it is necessary to reverse the natural or free flow direction of a fluid. For example, in water supply systems, water stored in reservoirs needs to be transported through pipes and raised to a higher pressure at the point of use. This requires the use of pumps and the input of mechanical energy to overcome friction losses and to raise the water pressure. Also, pressurised air is needed in a pneumatic tyre, and this requires mechanically driven compressors.

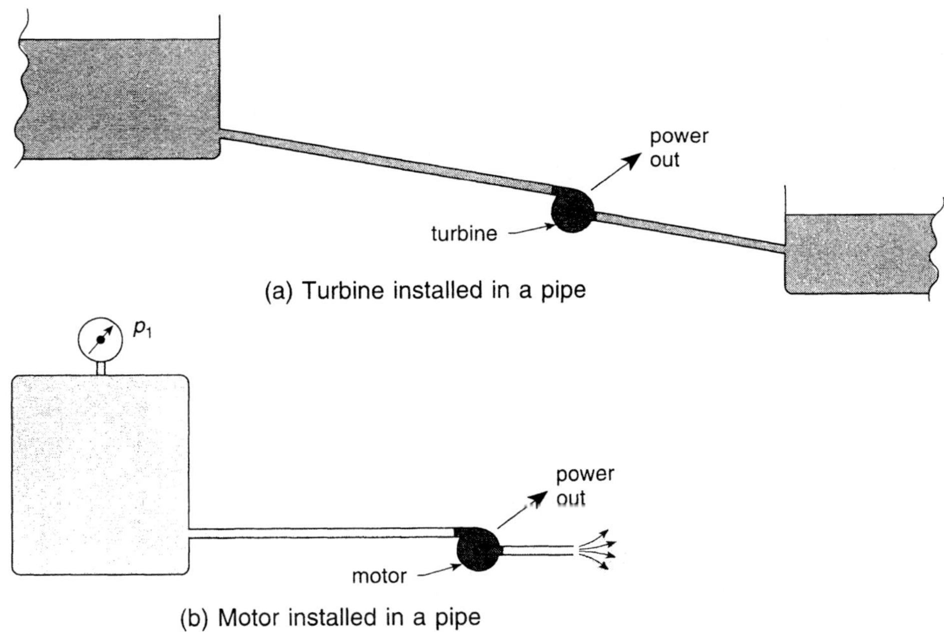

Fig. 5.2 *Using the free flow of a fluid to produce mechanical power*

In this chapter the most common types of fluid machinery are discussed and their operating characteristics considered.

5.1 *BASIC CONCEPTS*

Before discussing the various types of fluid machinery, it is necessary to have an understanding of some of the basic concepts and terminology as follows:

Rotary motion: Circular motion around an axle or shaft.

Linear motion: Motion in a straight line.

Reciprocating motion: Motion which alternates between one direction and the reverse of that direction. In fluid machinery, reciprocating motion usually occurs as reciprocating linear motion. Usually there is a piston or diaphragm connected to a rotating crankshaft by means of a connecting rod or cam mechanism. The crankshaft rotates continually in one direction but the piston or diaphragm moves with reciprocating linear motion.

Reversal of motion is often rapid, for example if a reciprocating machine is driven by a crankshaft rotating at 3000 rpm there are 50 cycles per second, and the reversal of motion occurs 100 times per second!

Positive-displacement fluid machinery: Fluid is entrained and held captive in an essentially leakproof chamber while the transfer of power from the fluid to the machine (or vice versa) occurs. It is usually necessary to provide some clearance between the moving and stationary components and in such cases, leakage (or backflow) can occur. The greater the pressure difference and the slower the movement, the greater is the leakage. In some cases the leakage can be negligible, for example in high-speed reciprocating gas compressors where the piston is a close fit in the cylinder and is oil lubricated. If a flexible diaphragm is used instead of a piston there is no leakage because no clearance is necessary.

Positive-displacement fluid machines are best suited to applications requiring a relatively high pressure (high head) but a relatively low flow rate.

Rotodynamic fluid machinery: The transfer of power from fluid to machine (or vice versa) is caused by the rotation of an impeller. The term rotodynamic is descriptive because the impeller rotates, and the power transfer occurs as a result of dynamic pressure forces acting on the impeller (that is, forces resulting from the motion of the impeller). A simple example of a rotodynamic fluid machine is the familiar household fan, where an electric motor rotates a shaft to which the impeller (a fan) is attached. Rotation of the impeller causes air to flow through the fan.

Rotodynamic fluid machines are best suited to applications requiring a relatively high flow rate but a relatively low pressure (low head).

Parallel and series arrangements: In many cases, fluid machines have a number of pistons or impellers connected to a common shaft. These can be arranged in parallel or series. In a *parallel arrangement* the inlet ports are in parallel and the outlet ports are in parallel. This increases the overall flow rate but does not change the overall head (pressure). In reciprocating machines the various pistons can be connected out of phase with one another, so this smooths the flow (reduces the impulsiveness). In a *series arrangement* the outlet port of the first stage is connected to the inlet port of the next stage, and so on. This does not change the overall flow rate, but increases the overall head (pressure). Connection in series is also known as *staging*. Usually the machines are designed so that the change of fluid head is the same in each stage.

Parallel and series arrangements for positive-displacement and rotodynamic pumps (or compressors) are illustrated in Figure 5.3. For simplicity, two cylinders or stages are shown, but in practice as many as ten may be used.

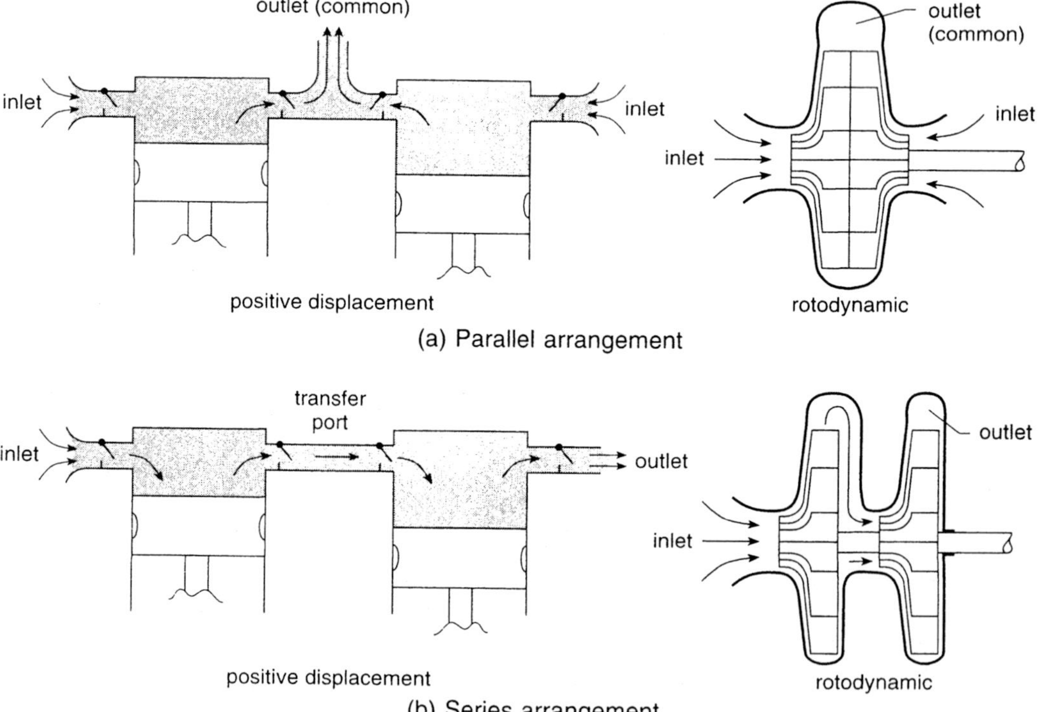

Fig. 5.3   *Parallel and series arrangements of fluid machinery*

5.2 *TYPES OF FLUID MACHINERY*

Fluid machinery can be divided into two basic groups, depending on the mode of operation:

- *Machines that convert mechanical power (energy) to fluid power (energy)*

  These machines are pumps, compressors, fans and propellers. Because mechanical power is converted to fluid power, the total head of the fluid is increased. That is, the total head of the fluid leaving the machine is greater than the total head of the fluid entering the machine. The mechanical power input to these machines is usually rotational and provided by an electric motor. Where portability is a prime consideration, the power input may be provided by a heat engine. For low-power applications, human muscle power may be used (for example a hand pump or a bicycle pump).

- *Machines that convert fluid power (energy) to mechanical power (energy)*

  These machines are turbines, motors or actuators. Because fluid power is converted to mechanical power, the total head of the fluid is less at outlet than it was at inlet. The mechanical power output of these machines is usually in the form of rotational motion but it may also be linear or reciprocating.

  Notes
  - The two basic types of fluid machinery are often connected in tandem in a single machine. For example, in a hydraulic jack, mechanical power is first converted into fluid power, and then the fluid power is reconverted to mechanical power. In a free-piston gas compressor, fluid power is first converted into mechanical power and then the mechanical power is reconverted into fluid power. This tandem coupling of fluid machinery is not treated in this chapter. Each type of machine is considered as a separate unit.
  - Some types of fluid machinery can operate in either of the two modes. For example, some pumps can function as turbines or motors when the fluid flows in the reverse direction. However, often it is undesirable for reverse flow to occur and in such cases one-way valves (or check valves) are incorporated internally or externally to prevent reverse flow. A pump or compressor with valves of this type will not operate as a turbine or motor in reverse (unless the valve arrangement is changed). Other types of fluid machines, such as Pelton-wheel turbines (impulse turbines), will not operate in reverse because of their inherent design.

  Within the two basic groups of fluid machinery there are many variations, as shown in Figure 5.4 on page 110.

  The choice of the most suitable type of machine for a given application depends on a number of factors:

  - Operating mode and whether or not reverse operation is required
  - Required pressure or head
  - Required volume flow rate
  - Nature of the fluid (liquid, gas or slurry)
  - Temperature of the fluid
  - Whether or not self-priming is needed (with liquids)
  - The environment in which the machine will operate and the available space

- Vibration and noise associated with the machine
- Operating speed and available power source
- Convenience of installation and use, including safety aspects
- Lubrication required and whether lubricant and fluid will mix
- Reliability
- Durability (life expectancy)
- Maintenance required over the operating life
- Efficiency
- Cost (both first cost and running cost).

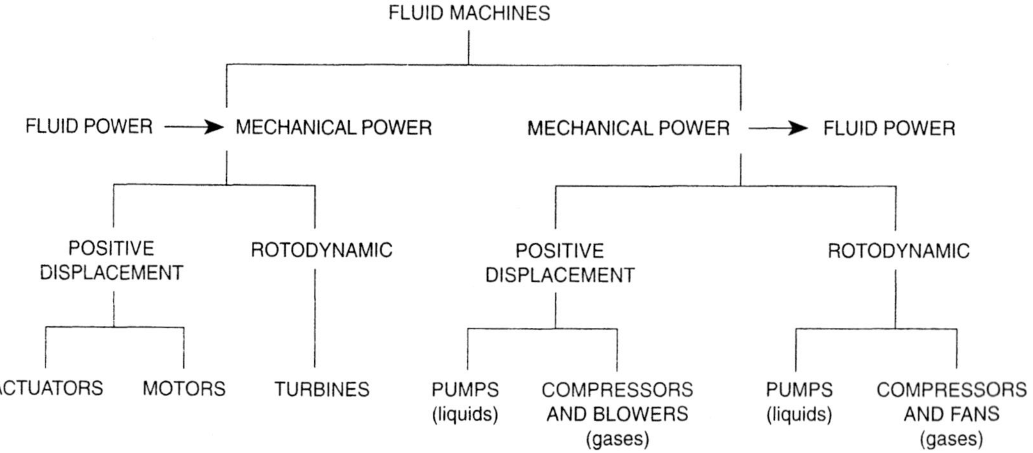

Fig. 5.4   *Types of fluid machinery*

The characteristics of the various types of fluid machines are now briefly outlined.

Pumps

In the broadest sense a pump can be defined as a device *that moves, compresses or exhausts a fluid.* Engineers usually define a pump as *a machine that converts mechanical power to fluid power.* In traditional engineering usage the word pump is usually used when the fluid is a liquid. When the fluid is a gas the fluid machine is usually called a compressor or fan[1].

Pumps are usually designed to transfer liquids from one location to another along a pipe, often to a higher elevation or pressure. To do this the pump increases the pressure of the liquid so that the pressure at the pump outlet is higher than the pressure at the pump inlet. That is, most of the head increase between the inlet and outlet of the pump is in the form of pressure head and this is illustrated in Example 5.1. This statement appears intuitively wrong as it seems logical that if a pump causes fluid to flow, the pump would need to substantially increase the velocity of the fluid. That is, it would seem that most of the head imparted by the pump would need to be in the form of velocity head. In fact, if the pump outlet and inlet ports have the same diameter, there is no velocity head increase at all between the inlet and outlet flanges of the pump. In some pumps the outlet port is smaller than the inlet port and then there is an increase in velocity head, but the increase in velocity head is usually only a small part of the total head increase.

---

[1] There are some exceptions, for example, *vacuum pumps* and *bicycle pumps*, which are used with air.

Notes

- The pressure at the outlet of a pump depends on the system resistance. That is, if liquid flows out of a pump to the atmosphere there is no resistance at the outlet of the pump and the pressure at the pump outlet is atmospheric. A pump increases the pressure of a fluid only when there is resistance at the outlet of the pump; that is, when the pump is delivering fluid against resistive forces (the system head at outlet).
- The maximum pressure the pump can deliver (the maximum pumping head) depends on the the design of the pump and the size and operating speed. For any given pump operating at constant speed the outlet pressure (and total head) are dependent on the flow rate. Usually the outlet pressure and pumping head are at maximum when the flow rate is small or zero.
- If there is resistance at the inlet side of the system, the inlet pressure at the pump can be reduced below atmospheric (vacuum). A pump for use with gases that produces a high vacuum at inlet is known as a vacuum pump. However, this is a special case, as most pumps are used with liquids, and then it is undesirable for the inlet pressure to be too low because the pump can cavitate. Cavitation is discussed later in this chapter.
- Usually the potential head increase across a pump is small, and in the case of an in-line pump (one in which the inlet and outlet ports are at the same horizontal level) there is no potential head difference across the pump at all. Also, if an in-line pump has inlet and outlet ports of the same diameter, there is no velocity head difference across the pump. In this case *all* the head developed by the pump is in the form of pressure head.
- If the inlet and outlet ports have the same diameter, there is no velocity head difference between the inlet and outlet flanges of the pump. However, there is usually a substantial increase in fluid velocity *within* the pump itself. The increased velocity is then reduced again before the fluid leaves the pump, and velocity head is converted to pressure head (as required by the Bernoulli equation).

Example 5.1

A centrifugal pump as illustrated in Figure 5.5 delivers 25 L/s of water with a total head of 15 m. Determine for this pump the following heads and their percentage of the total head:

(a) potential head;
(b) velocity head;
(c) pressure head.

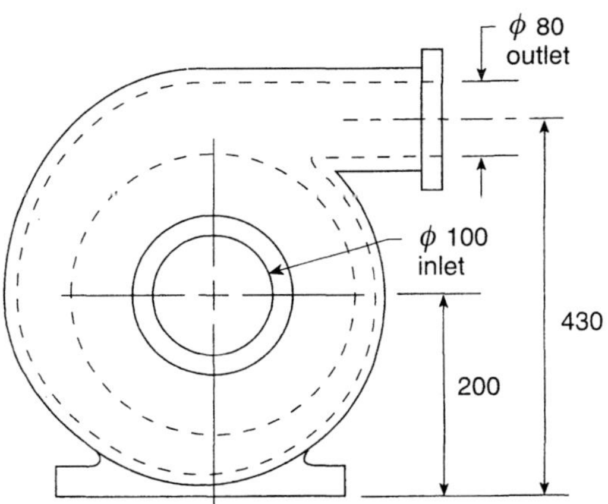

Fig. 5.5

Solution

(a) The potential head is $430 - 200 = 230$ mm $= \mathbf{0.23}$ **m**

% of the total head is: $\dfrac{0.23}{15} \times 100 = \mathbf{1.53\%}$

(b) The velocity at inlet is: $\dfrac{0.025}{A(0.1)} = 3.183$ m/s

The velocity at outlet is: $\dfrac{0.025}{A(0.08)} = 4.974$ m/s

The velocity head is: $\dfrac{4.974^2 - 3.183^2}{19.62} = \mathbf{0.744}$ **m**

% of the total head is: $\dfrac{0.744}{15} \times 100 = \mathbf{4.96\%}$

(c) The pressure head is: $15 - 0.23 - 0.744 = \mathbf{14.03}$ **m**

% of the total head is: $\dfrac{14.03}{15} \times 100 = \mathbf{93.5\%}$

Note   In this pump, there is a small increase in velocity head and potential head, but the increase in pressure head is by far the greater proportion of the total head.

Types of pump

Pumps can be classified in two main groups:

- **Positive-displacement pumps.**   These include piston pumps, diaphragm pumps, gear pumps, lobe pumps, geroter pumps, vane pumps, screw pumps and peristaltic pumps. Most positive-displacement pumps are fixed-volume pumps because the volumetric displacement cannot be changed at a given pump speed. However, there are some ingenious designs, such as swashplate, bent-axis and radial-piston pumps, which are classed as variable-volume pumps because the volumetric displacement is variable and can be adjusted either manually or automatically in response to the outlet pressure.

- **Rotodynamic pumps.**   These types of pump fall into one of three classes: axial-flow, radial-flow and mixed-flow types, depending on the direction of the fluid leaving the impeller.

Advantages and disadvantages of positive-displacement pumps compared to rotodynamic pumps

Because there are so many types of positive-displacement and rotodynamic pumps it is difficult to generalise about their relative advantages and disadvantages. However, the following is a guide to the advantages and disadvantages of positive-displacement pumps, compared to rotodynamic pumps.

Advantages

- High pressures (and heads) can be developed with a small, compact pump, whereas rotodynamic pumps cannot develop high pressures unless they have large impellers or many stages, and hence are much larger in size. For example, small, single-stage commercially available hydraulic positive-displacement pumps can deliver pressures in excess of 70 MPa, whereas much larger commercial rotodynamic pumps have a pressure limit of about 1.5 MPa for a single stage, and about 6 MPa for multistage.

- High efficiency (up to about 90%). Rotodynamic pumps usually have a lower efficiency under most operating conditions and in most cases it is below 80%.
- High flexibility, that is, they can operate efficiently over a wider range of speeds and pressures than is the case with rotodynamic pumps. That is, the efficiency of positive-displacement pumps does not change as much with flow rate and head, as is the case with rotodynamic pumps.
- Self-priming capability; that is, on start-up they do not need to be primed, and they can draw liquid up from a reservoir below the pump. This eliminates the need for check valves (which can leak over a prolonged time). Also, if any air or vapour gets into the line the pump will continue to operate and clear the air or vapour. Rotodynamic pumps are not inherently self-priming and will no longer pump liquid if an air or vapour lock occurs (the impeller is surrounded by gas).[2]
- Relatively constant flow rate is maintained despite large changes in system pressure (system head). This is sometimes an advantage and is not the case with rotodynamic pumps, whose flow rate is far more sensitive to system head.

Disadvantages

- They have a relatively low flow rate compared to rotodynamic pumps.
- Most types are not suitable for fluids containing abrasive particles because scoring and wear tend to occur in the fine clearances between the moving parts.
- They are generally more expensive.
- They have more moving parts and tend to need more maintenance than rotodynamic pumps.
- If the moving parts require lubrication by a fluid other than the fluid being pumped, and fluid is in contact with the moving parts, lubricant can mix with, and contaminate, the fluid. (The fluid can also contaminate the lubricant.)
- They usually produce more noise and vibration than rotodynamic pumps, which are much quieter and produce little vibration.
- The flow tends to be pulsating (especially reciprocating types) whereas rotodynamic pumps produce a smooth, even flow of fluid. Pulsating flow is undesirable, and there is more likelihood of cavitation when it occurs because of the rapid acceleration and deceleration of the fluid.
- The pump or system usually requires some form of pressure-limiting device (such as a pressure-relief valve) because if a blockage or valve closure occurs on the discharge side of the system the pressure continues to build up and can reach a dangerous level. This is not the case with rotodynamic pumps as they have a maximum pressure limit, even with a totally blocked discharge.

Compressors

A compressor is essentially a pump used with a gas rather than a liquid, in which the outlet of the pump is connected to a pressure vessel where the compressed gas accumulates. Compressors fall into the same two categories as pumps, that is, positive-displacement and rotodynamic types. The most common positive-displacement types are piston, vane and screw compressors, whereas the most common rotodynamic types are the centrifugal-flow and axial-flow compressors.

---

[2] There are some special designs of rotodynamic pumps which can retain some liquid from the initial prime, and this enables them to continue to pump liquid should a spasmodic air or vapour lock occur.

Fans

Fans move large quantities of gas without achieving a significant pressure increase. That is, they deliver large volumes of gas with a small increase in pressure (head). They are often rotodynamic, the most common being the axial-flow type, although centrifugal types are common.

Blowers

Blowers are similar to fans except that there is usually a greater pressure increase. For example, blowers are often used with furnaces or internal-combustion engines to increase the flow of air into the combustion chamber. Positive-displacement blowers are common and usually are of the lobe type.

Propellers

A propeller is a rotating impeller with helical blades. As the name suggests, they can obtain a large reaction thrust force to propel a boat or aircraft. This force results from the momentum change of the fluid as it passes over the propeller. Propellers are also used as axial-flow pump impellers and as wind-turbine impellers.

Stirrers

Stirrers are used to mix or agitate a fluid and are widely used in applications such as paint mixing and food processing. Stirrers usually have impellers which may be small propellers. In this case axial thrust is undesirable. This thrust may be minimised by mounting two impellers with opposite blade directions on a common shaft.

Fluid motors and turbines

Fluid motors and turbines operate in the reverse mode to pumps or compressors, that is, they convert fluid power to mechanical power. The function of fluid motors and turbines is the same but in engineering the word *motor* is usually used with positive-displacement types and the word *turbine* with rotodynamic types.

Actuators

Like fluid motors and turbines, actuators convert fluid power to mechanical power. In the broadest sense, fluid motors and turbines can be classed as rotary actuators. However, in engineering, the word *actuator* is usually used to mean fluid machinery of the reciprocating type, in particular, pneumatic and hydraulic cylinders. Pneumatic and hydraulic cylinders are cylinders fitted with a close-fitting piston which is driven by hydraulic or pneumatic pressure. A rod or other attachment on the piston transmits the piston force to a mechanism or machine. The piston may be powered in both forward and return directions (double-acting) or powered forward, with a spring-loaded return (single-acting).

Notes

- Actuators can produce a reciprocating rotary motion, but these types are less common than the linear motion types.

- Rodless actuators are available for pneumatic applications. They have a sliding seal at the side of the cylinder, and the cylinder force is transmitted through a pin or spigot attached to the side of the piston. Alternatively, the piston may contain a strong permanent magnet that transmits force to an external guide ring of magnetic material. Rodless actuators are more compact than the rod type, and buckling of the rod in compression is avoided. However, they are also more expensive.
- Pneumatic actuators of a bellows design using neoprene are commercially available. Their stroke is limited to about 350 mm but they are much cheaper than actuators of the piston-in-cylinder type.

Self-test problem 5.1

The hydraulic cylinder illustrated in Figure 5.6 is supplied with oil at a pressure of 2.5 MPa. If frictional losses are neglected, determine:
(a) the force F exerted by the rod on the forward stroke;
(b) the force F exerted by the rod on the return stroke;
(c) the flow rate of oil to the cylinder on the forward stroke in L/s if the velocity of the rod on the forward stroke is 0.2 m/s;
(d) the mechanical power developed during the forward stroke;
(e) the fluid power developed during the forward stroke.

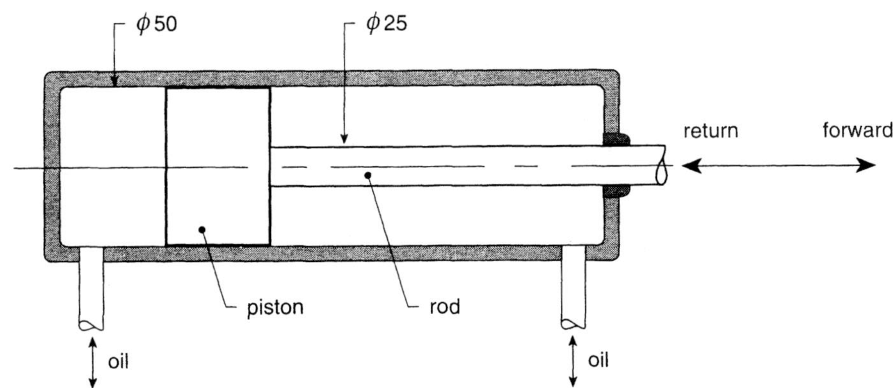

Fig. 5.6

5.3 *TYPES OF POSITIVE-DISPLACEMENT PUMP*

A large number of different types of positive-displacement pump are manufactured. The most common types are now described.

Piston pump

The piston pump, as illustrated in Figure 5.7 on page 116, has a piston which reciprocates in a cylinder. If driven by a rotating shaft, it is necessary to convert the rotary motion to a reciprocating motion. This is accomplished by a mechanism such as a crankshaft and connecting rod, scotch yoke, or cam. One-way inlet and outlet valves are located at the top of the cylinder. These are often spring loaded or of the reed-valve type, so they are self-actuating (no camshaft or mechanism is necessary to open and close them). In some cases mechanically driven poppet valves, or a rotary valve plate, are used. The clearance between the piston and the cylinder is small, and lubricant is required in order to reduce wear and leakage. Consequently these pumps are usually provided with filters to remove contaminants before the fluid enters the pump.

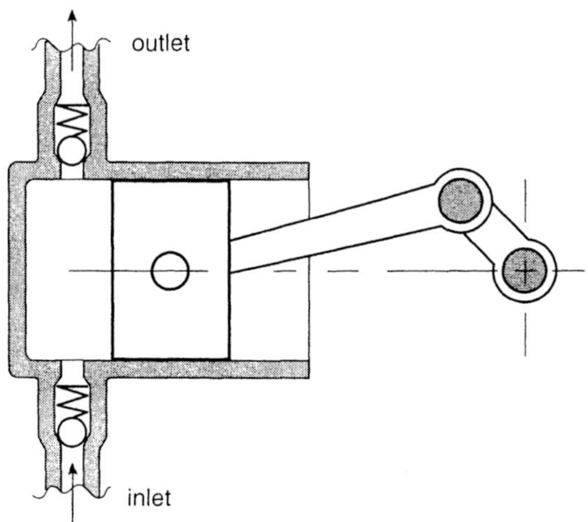

Fig 5.7   *Reciprocating piston pump*

Reciprocating piston pumps often have more than one cylinder. Multicylinder configurations have the advantage of smaller pistons (for a given flow rate) and thus higher pressures are possible with smoother motor torque. Also flow is less pulsating because the various pistons are out of phase with one another. Many configurations are possible, including in-line, opposed, axially-displaced pistons, and radially-displaced pistons.

Most reciprocating piston pumps of the type illustrated in Figure 5.7 have a fixed volumetric displacement. This means that if the pump is driven by a motor which runs at constant speed, the flow cannot be controlled simply by the use of valves in the system. For example, if a valve in the discharge line is partially closed, the result will not be a reduction in the flow rate but an increase in the pressure at outlet. A pressure-relief valve is therefore necessary to prevent damage to the pump or the system as a result of excessive pressure. Also, many systems use on-off control with a pressure switch to shut off the power supply to the pump as soon as the system pressure reaches the set-point limit.

Some multicylinder reciprocating pumps can deliver variable volume at constant speed. In these pumps the volume is controlled by varying the stroke of the piston. This control can be achieved manually with a handwheel, or automatically by a pressure-actuated mechanism. In the automatic system, as the pressure increases, the stroke reduces, until eventually the pump will deliver no fluid at all when the set-point pressure is reached. However, variable-volume reciprocating pumps are expensive and therefore usually they are used only in fluid-power applications such as sophisticated hydraulic systems.

Common types of variable-volume, multicylinder reciprocating pumps are the bent-axis, swashplate and radial-piston types. The principle of the bent-axis type is illustrated in Figure 5.8. The offset angle is variable and as it reduces, the stroke and delivery reduces, until when the offset angle is zero, there is no delivery at all.

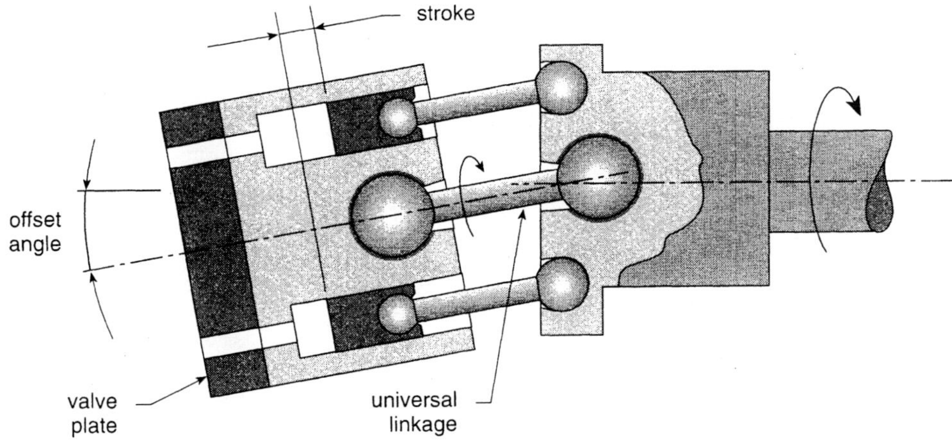

Fig. 5.8 *Operating principle of a bent-axis variable-volume reciprocating piston pump*

Piston pumps can achieve high pressures and are often used in hydraulic systems where the hydraulic oil also serves as a lubricant. Piston pumps are also used for manually operated pumps for oil or water and for air (as in a bicycle pump). As noted previously, piston pumps used with gases are more correctly called compressors (or vacuum pumps).

Plunger pump

The plunger pump illustrated in Figure 5.9 is essentially a reciprocating pump where a plunger replaces the piston. The difference between a plunger and a piston is that a plunger is not a close fit in a bore, and considerable clearance occurs around it. A seal on the plunger is provided. This eliminates the need for lubrication between the piston and the bore and eliminates the scoring problems between them that can occur if the fluid contains abrasive particles.

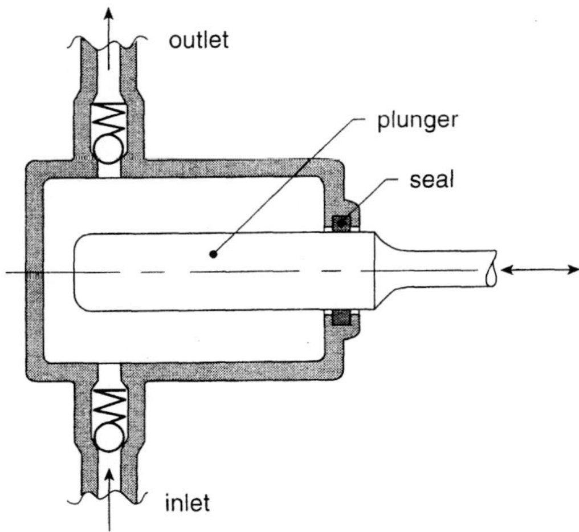

Fig. 5.9 *Plunger pump*

Diaphragm pump

The diaphragm pump illustrated in Figure 5.10 is essentially a reciprocating pump where a diaphragm made of a flexible material such as neoprene replaces the metallic piston. This eliminates the need for fine clearances, lubrication between piston and cylinder, and the possibility of leakage (unless the diaphragm is punctured). Diaphragm pumps can pump a wide variety of fluids including liquids containing solid matter held in suspension, such as slurries or dusty air. They are relatively inexpensive but because of the flexible diaphragm they cannot achieve high pressures, or deal with high-temperature fluids.

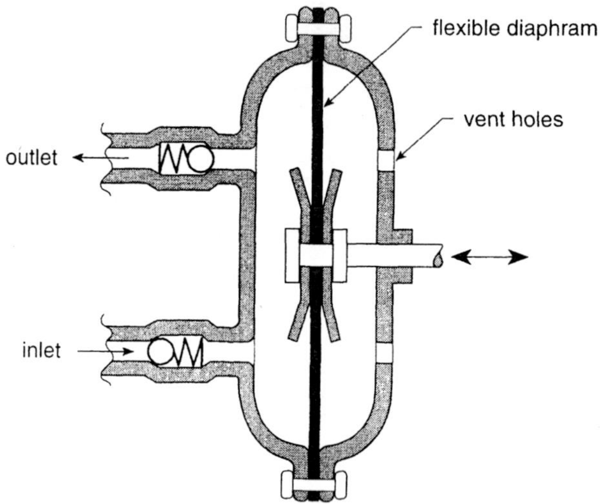

Fig. 5.10   *Diaphragm pump*

Because of their low cost, self-priming capability, zero leakage and compatibility with a wide variety of fluids, diaphragm pumps are widely used for a diverse range of applications, such as muddy-water drainage pumps, petrol pumps in motor vehicles, and water pumps in caravans and boats.

Single- and double-acting reciprocating pumps

In reciprocating piston or diaphragm pumps the fluid is usually in contact with only one side of the piston or diaphragm. This is known as a single-acting pump. However, it is possible to use both sides of the piston or diaphragm and this is known as a double-acting pump. Both arrangements are illustrated in Figure 5.11 (opposite).

The double-acting arrangement is more compact because, in effect, one double-acting piston or diaphragm is equivalent to two single-acting ones. Also there is a continuous delivery of fluid, whereas in the single-acting type, delivery occurs for only one-half of the cycle (for a single piston or diaphragm). However, sealing and lubrication are more difficult, and the mechanical arrangement is more complex, so double-acting pumps are generally less common than single-acting ones.

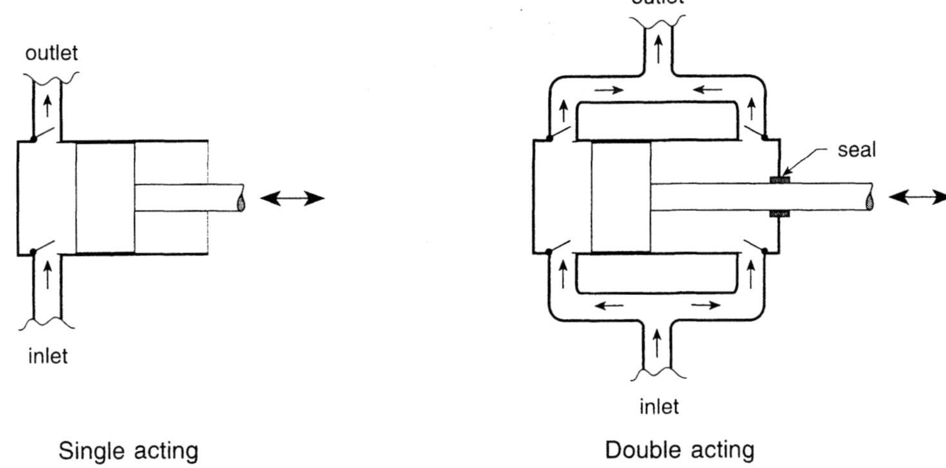

Fig. 5.11　*Single- and double-acting pumps*

Gear pumps

In gear pumps, as illustrated in Figure 5.12, one gear is driven by a motor, whereas the other is an idler gear and rotates because of the meshing action between the two gears. Usually spur gears are used, but helical and herringbone gears are sometimes used. Gear pumps may be of the external or internal gear type. In the *external gear* type, the fluid is carried around the external teeth in both gears. In the *internal gear* type, the fluid is carried around external teeth in the inner gear, which is driven, and also around internal teeth in the outer gear, which is the idler gear. A crescent between the inner and outer gears acts as a seal between the inlet and outlet sides of the pump. This type is more compact than the external gear type for the same flow rate.

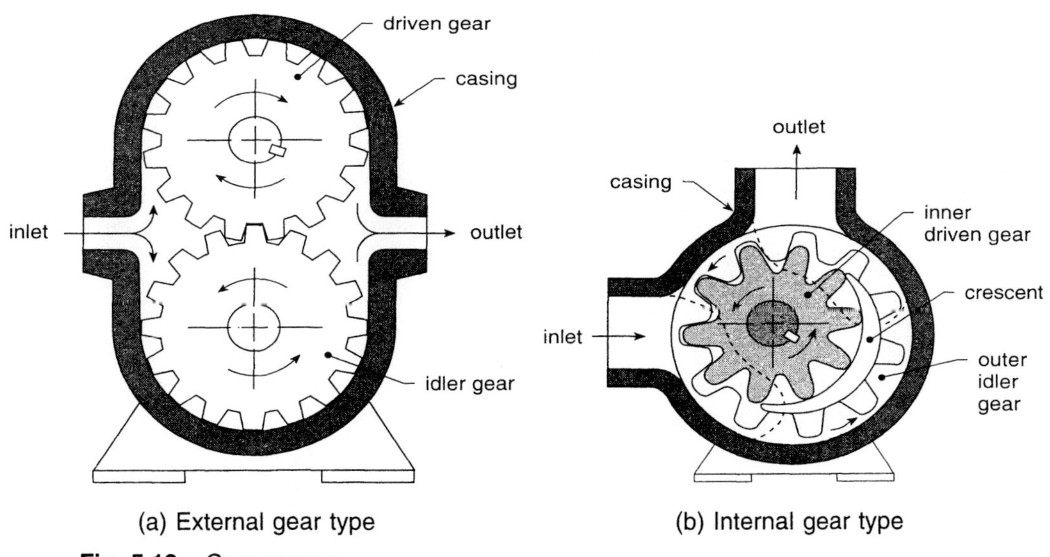

Fig. 5.12　*Gear pumps*

Unlike in reciprocating pumps, valves are not necessary for pumping action to occur, and little pulsation occurs in the flow. However, fine clearances are necessary around the sides of the gears and between the gears and the housing, so at high pressures some internal backflow occurs. Gear pumps are relatively inexpensive, but can be a source of noise at high speed, due to the meshing of the gears. They are used for intermediate pressure applications such as oil pressure pumps for internal combustion engines.

Lobe pump

The lobe pump illustrated in Figure 5.13 operates in a similar fashion to an external gear pump. The fluid is trapped between the lobes and the casing, and is carried around from inlet to outlet. Unlike the gear pump, *both* lobes are externally driven (usually by a motor driving mating gears) so there is no contact at all between the lobes, and no wear should occur between the lobe faces. Like the gear pump, side clearance is necessary, so lobe pumps tend to have more internal backflow than gear pumps (for the same pressure difference) because there is also internal clearance between the lobes. Hence lobe pumps are not suitable where high pressures are required.

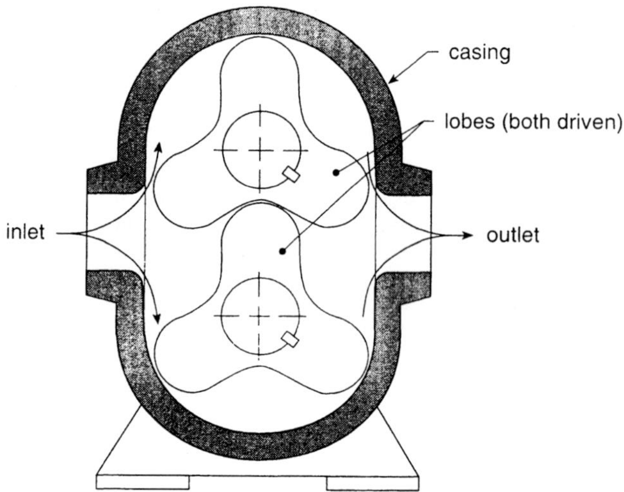

Fig. 5.13 *Lobe pump*

Their volumetric displacement is greater than gear pumps (because of the greater lobe volume) but flow is also more pulsating. Lobe pumps are often used with gases as blowers, for example in two-stroke diesel engines or other engines which have a forced air supply. They are also used for low-pressure liquid pumping applications.

Geroter pump

The geroter pump, illustrated in Figure 5.14 (opposite), is somewhat of a hybrid between an internal gear pump and a lobe pump. Like the internal gear pump, the inner geroter rotates and carries the outer geroter around with it. However, the inner geroter has one less lobe than the outer one so the lobes go into and out of mesh as they rotate. The inlet port is located where the lobes move out of mesh and the outlet port is located where the lobes move into mesh.

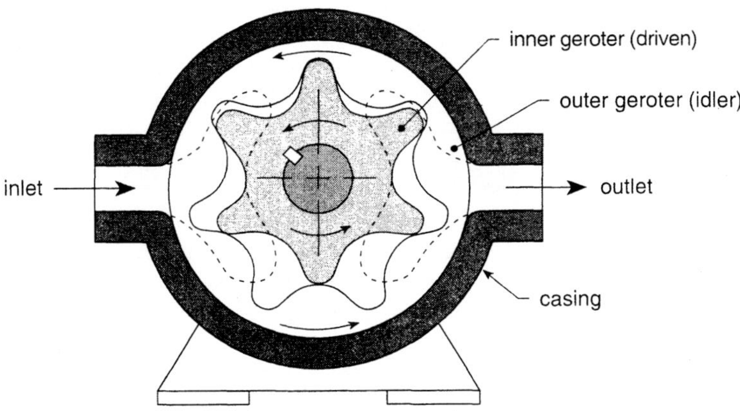

Fig. 5.14 *Geroter pump*

Like gear pumps, geroter pumps are used for intermediate pressure applications such as oil pumps.

Vane pump

The vane pump, illustrated in Figure 5.15, has a slotted rotor into which sliding vanes are fitted. The housing into which the rotor is fitted is eccentric relative to the rotor. As the rotor rotates, centrifugal force drives the vanes outward to effect a seal between the vane and the casing. The inlet port is located in the space where the volume between the vanes increases, and the outlet port is located in the space where the volume decreases.

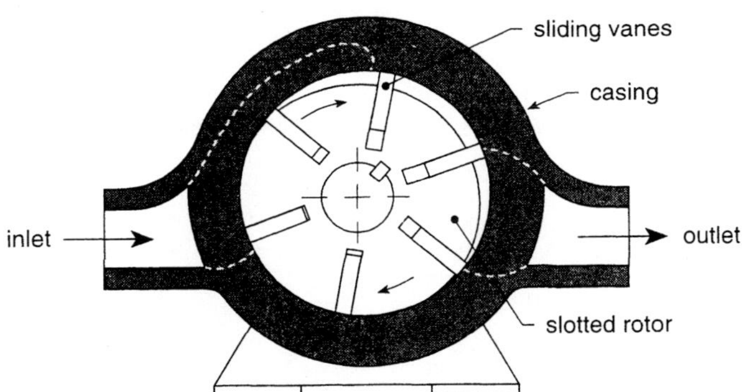

Fig. 5.15 *Vane pump*

Fine clearances are necessary between the vanes and the rotor, and also between the sides of the vanes, rotor and casing. Oil lubrication is necessary, and rapid wear will occur if the pumps run dry. Vane pumps are used for hydraulic applications, and also as rotary gas compressors. Vane pumps have several variations:

- In the balanced-vane pump the casing is elliptical in shape and there are two inlet and outlet ports located opposite one another. As the name suggests, in this pump, side load on the bearings due to the pressure difference between inlet and outlet is eliminated.

- In the variable-displacement or pressure-compensated vane pump, there is an outer ring held in the casing by a spring-loaded block and a hydraulic piston. As the system pressure increases, the hydraulic piston moves the outer ring towards the spring, and this reduces the eccentricity and discharge. When design pressure is reached, the outer ring is concentric with the rotor, so the delivery is zero. However, this design cannot be hydraulically balanced.
- In the flexible rotor type, illustrated in Figure 5.16, the rotor and sliding vanes are replaced by a star-shaped rotor made of a flexible material. The vanes in the rotor deform and flex, and thus provide the sealing effect of the sliding vanes. These pumps are much less costly than the sliding-vane type and the lubricant does not necessarily need to be oil. They are often used as water pumps in outboard motors and for similar applications. In this case, water acts as the lubricant, and rapid wear will occur if the pump is run dry, or if solid particles such as sand are drawn up into the pump.

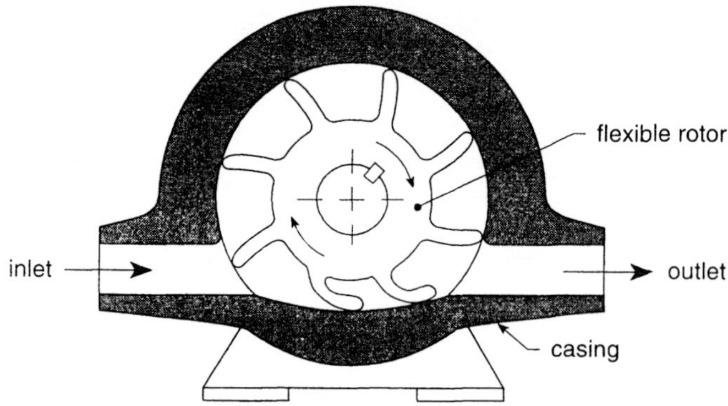

Fig. 5.16   *Valve pump—flexible rotor type*

Screw pump

In a screw pump, one or more rotors are in the form of a screw. In the three-rotor type illustrated in Figure 5.17 there is a centre-driven rotor and two outer idler rotors that mesh with it. Fluid enters axially and is conveyed axially from inlet to outlet, due to the meshing action between the rotors.

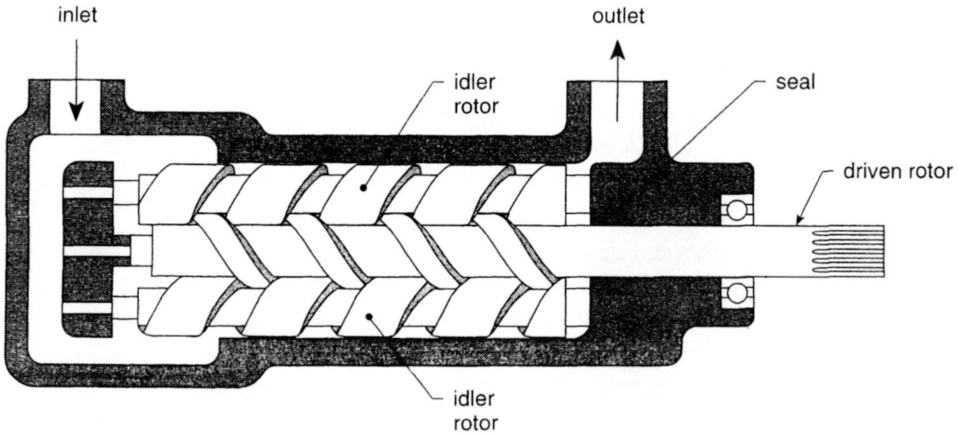

Fig. 5.17   *Screw pump (untimed triple-rotor, single-end type)*

Screw pumps have many variations[3] depending on whether:

- there are one or more rotors;
- the rotors are timed or untimed;
- the rotors are single- or double-ended.

Number of rotors: In the single-rotor type, the stator is made of a resilient material in the form of a double internal helix. The helical rotor is made of hardened steel and it presses on the flexible stator to maintain a seal. The cavity between rotor and stator forms a chamber which moves progressively along the helix as the rotor rotates, thus conveying the fluid from inlet to outlet. In the double-rotor type, there are two meshing rotors. In the triple-rotor type, there are three meshing rotors. This type is illustrated in Figure 5.17.

Timed or untimed rotor types: In the timed-rotor type, the rotors are supported at both ends by bearings and driven by meshing gears. There is a fine clearance between the rotors so there is no actual contact between them. In the untimed-rotor type the idler rotors are supported by the housing which also acts as a bearing. The driven rotor drives the idler rotors and there are no gears between them. This type is illustrated in Figure 5.17.

Single- or double-ended types: In the single-ended type, as illustrated in Figure 5.17, the fluid enters at one end and leaves at the other. In the double-ended type, each rotor is in the form of a half-left-hand and half-right-hand screw. The fluid enters at the centre and is conveyed to each end, where it leaves the rotor and flows to a common outlet. This has the advantage that the thrust forces on the rotor are internally balanced.

Screw pumps are widely used for pumping a variety of fluids, from low viscosity fluids such as water or petrol, to high viscosity fluids such as heavy oil or molasses. They are also used for conveying and compressing plastic pellets, and molten plastic for plastics extrusion and injection moulding. When the fluid is a gas the screw pump becomes a gas compressor and these are often preferred to the piston or vane types because they are quieter, produce less vibration and pulsation, and require very little maintenance.

The main disadvantage of screw pumps and screw compressors is their relatively high cost.

Peristaltic pump

In all the pumps thus far considered the fluid is in contact with the mechanical parts of the pump. In the peristaltic or flexible-tube pump, illustrated in Figure 5.18 on page 124, this contact is eliminated. The fluid passes through a tube and is squeezed from inlet to outlet by rollers attached to a rotating arm.

Peristaltic pumps are not very efficient, and are not suitable for high pressures or flow rates, or for fluids which are at a high temperature. However, they are widely used in the food, drug and medical industries because of the hygienic, contamination-free nature of their operation. Usually the tube is a disposable item, in which case no cleaning at all is required.

Peristaltic pumps are simple, inexpensive and reliable, and can be used for fluids difficult to pump by any other method, such as blood or cement slurry (concrete).

---

[3] The earliest type of screw pump was the Archimedean screw pump. This pump was used for lifting water. It cannot, however, develop pressure, and hence is now seldom used.

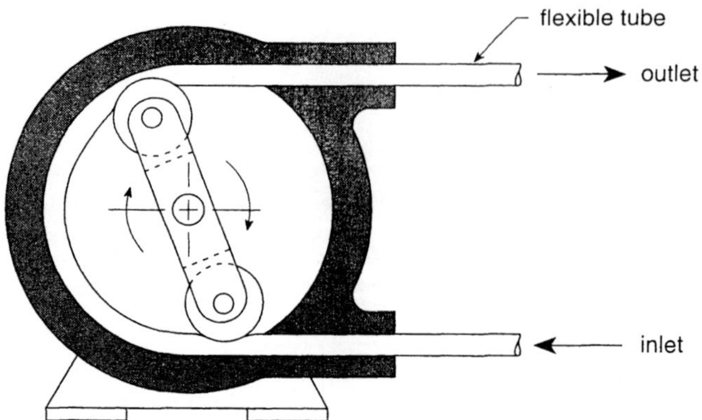

Fig. 5.18   *Peristaltic pump*

5.4   *TYPES OF ROTODYNAMIC PUMP*

There are three main types of rotodynamic pump: axial, radial or mixed flow. These types are classified according to the flow direction of the fluid *leaving* the impeller, and are illustrated in Figure 5.19.

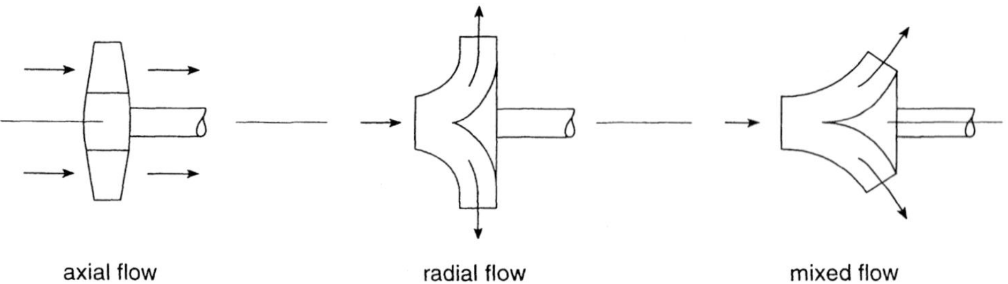

Fig. 5.19   *Types of rotodynamic pump*

Axial-flow or propeller pump: The flow direction at outlet is parallel to the impeller axis. That is, the fluid enters the impeller axially and leaves axially.

Radial-flow or centrifugal pump: The flow direction at outlet is radial (or perpendicular) to the impeller axis. That is, the fluid enters the impeller axially and leaves radially.

Mixed-flow pump: The flow direction at outlet is at some angle intermediate between parallel and perpendicular to the impeller axis. That is, the fluid enters the impeller axially and leaves at some angle intermediate between 0 and 90°.

All rotodynamic pumps depend on a rotating impeller to change the velocity and pressure of the fluid as it flows through the impeller. There is considerable clearance between the rotating impeller and the stationary parts of the pump, so rotodynamic pumps cannot produce the high pressures (heads) of positive-displacement pumps. Also they are not self-priming. However, they have a high volume flow rate and many other advantages over positive-displacement pumps, which accounts for their widespread usage.

The three main types of rotodynamic pump are now briefly described.

Axial-flow or propeller pumps

As shown in Figure 5.20, the impeller of these pumps is shaped like a propeller or fan. A number of blades, usually between three and six, are inclined at an angle to the rotating shaft. For maximum efficiency this angle varies, and becomes smaller, the greater the distance from the shaft centre line. Also, for high efficiency the blades should have an aerofoil shape to reduce turbulence and promote smooth passage of fluid.

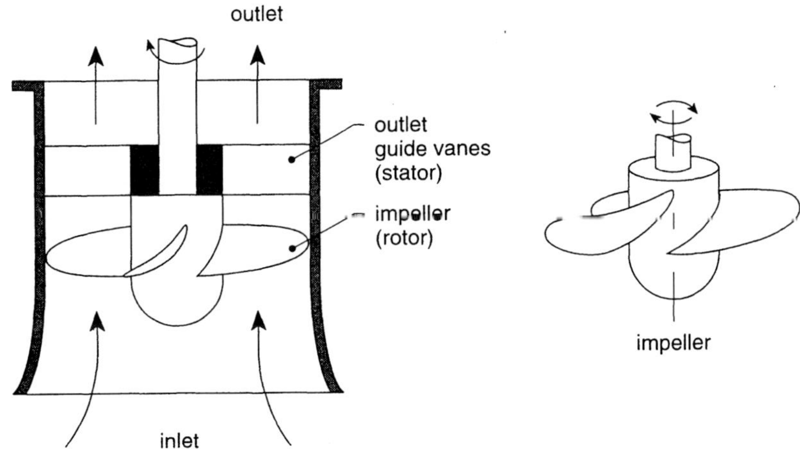

Fig. 5.20　*Axial-flow pump*

Axial-flow pumps are designed to move the fluid in an axial direction only but in practice this is difficult to achieve for two reasons:

- The rotation of the impeller imparts a rotary motion to the fluid. That is, the fluid tends to follow a spiral path from inlet to outlet. This effect can be minimised by fitting inlet or outlet guide vanes (or both) to redirect the fluid to an axial path. This set of vanes is also known as the stator. The axial-flow pump illustrated in Figure 5.20 has been fitted with outlet guide vanes for this reason.
- The rotary motion of the fluid induces a centrifugal force which tends to throw the fluid outward. This effect can be minimised by shrouding the impeller, or fitting it inside a housing so that the fluid is prevented from flowing outward. The axial-flow pump illustrated in Figure 5.20 has an impeller fitted into a housing so that most of the radial flow is prevented.

Axial flow pumps develop a small head (or pressure) per stage, and the ratio of outlet to inlet head (or pressure) is usually only about 1.2 per stage. However, they can readily be staged, and with 8 stages, the overall head (or pressure) ratio increases to about 4. They are capable of moving large volumes of fluid and do not change the flow direction, so they are used when a large flow rate is required with a small head, that is, when the fluid is pumped through a small height or pressure increase. When the fluid is a gas, an axial-flow pump becomes a gas compressor and in the multistage version is widely used in aircraft turbo-jet engines.

Radial-flow or centrifugal pumps

As the name suggests, centrifugal pumps use centrifugal force to cause the fluid to move radially outward through the impeller. The fluid enters the impeller axially and is held in chambers formed by the impeller vanes. Rotation of the impeller produces a centrifugal force which drives the fluid radially outward from the centre of rotation. The fluid leaves the impeller with a higher velocity than it had at inlet, that is, the velocity head is increased. The velocity head of the fluid leaving the impeller is converted to pressure head by reducing the velocity of the fluid in a gradually expanding section around the periphery of the impeller, known as the volute. Alternatively, a series of stationary guide vanes, known as the diffuser, redirects the fluid and reduces its velocity. This is a more efficient but more costly method and the size of the pump is increased, so diffuser-type centrifugal pumps are less common than volute types.

A typical volute pump type with single suction is illustrated in Figure 5.21.

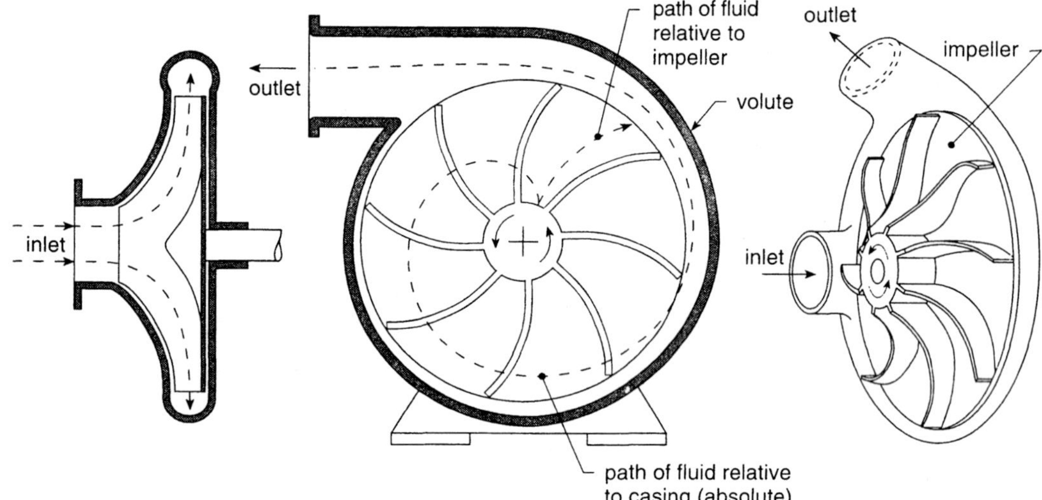

Fig. 5.21   *Centrifugal pump (single-suction, volute type)*

Notes

*   Note in Figure 5.21 that the blades of the impeller are inclined backward to the direction of rotation. This is the most common shape of the impeller blades, but radial (straight) blades or forward-curved blades are sometimes used.
*   The path of the fluid is shown relative to the impeller (relative velocity) and relative to the casing (absolute velocity). Relative to the impeller the fluid moves radially outward and backward, flowing through the passage formed by the blades. Relative to the casing the fluid moves forward in a spiral path because the absolute velocity is the vector sum of the radial component (caused by centrifugal force) and the tangential component (caused by the forward rotation of the impeller). The shape of the volute matches the direction of absolute velocity of the fluid at the periphery of the impeller, so the fluid flows smoothly from impeller to volute and through to the outlet of the pump.
*   An open-face impeller has been shown. This is a common design but some impellers are closed at the front face as well as the back face. This type is known as a shrouded

impeller. Shrouded impellers produce a higher efficiency than open impellers but their manufacturing cost is higher.

- A single-suction impeller has been shown. This is a common design, but for a greater flow rate, and in order to obtain a balanced axial force on the shaft, double-suction impellers (as illustrated in Figure 5.3—parallel arrangement) are sometimes used.
- The impeller is shown with eight blades, which is about the optimum number for good efficiency. An increase in the number of blades produces a better flow path with less circulatory flow, but increases the frictional loss, so it is important to optimise the number of blades for best efficiency. However, in small pumps when efficiency is not a critical factor, impellers with as few as two blades may sometimes be found.
- The head (and pressure) developed by a centrifugal pump varies in proportion to the impeller speed and size. The greater the speed and size, the greater is the head developed. For a given inlet pipe diameter, the size of the impeller depends on the ratio of the outside to inside diameter (*OD/ID* ratio) and the higher this ratio, the greater is the head.
- In Figure 5.21 a single-stage pump has been shown. Centrifugal pumps are sometimes made with staged impellers (as illustrated in Figure 5.3—series arrangement). Centrifugal pumps do not lend themselves to staging as well as axial-flow pumps because the fluid has to be directed back from the outward radial direction, inward and then to the axial direction for the next stage. However, staged centrifugal pumps are manufactured for special applications where a high head is required with a relatively small impeller diameter.
- Like all rotodynamic pumps, centrifugal pumps are not self-priming. Priming is not necessary if the pump inlet is below the liquid surface level but this is often not practical. One common solution is to make the pump submersible. Submersible pumps can be placed in the tank containing liquid and below the surface. This eliminates the priming problem and the need to have outlet pipes and valves fitted to the tank, but introduces corrosion, sealing and safety problems (when high-voltage electrical motors are used). However, submersible centrifugal pumps are commercially available and often used.

Centrifugal pumps are the most common of the rotodynamic pumps and are manufactured in a large range of designs, sizes, speeds, and materials. Small centrifugal pumps are used as water pumps in portable applications such as in caravans and boats. These pumps often have plastic impellers and housings. For industrial applications, standard impeller sizes range from about 100 mm to 500 mm with port sizes from about 32 mm to 300 mm. These pumps can produce heads of 150 m or more with a single-stage impeller pumping water. Typical impeller materials are cast iron, gunmetal, bronze or stainless steel.

Mixed-flow pumps

A mixed-flow pump is a hybrid between the axial-flow and centrifugal pumps and consequently has a performance intermediate between these two types. The impeller causes the fluid to flow in both an axial and a radial direction simultaneously. Two basic types are used: the divergent-cone type and the volute type. These are illustrated in Figure 5.22 on page 128.

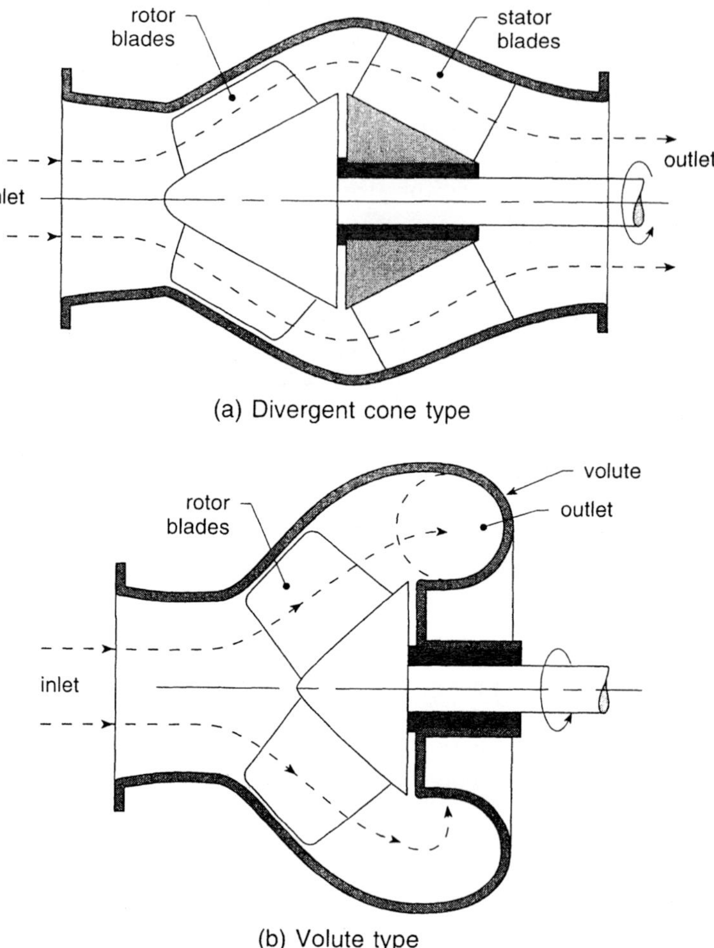

Fig. 5.22   *Mixed-flow pumps*

In the divergent-cone type, the fluid leaving the impeller is redirected through guide vanes back to an axial path (as an axial-flow pump). In the volute type, the fluid leaving the impeller flows through a volute and leaves the pump in a radial direction (as in a centrifugal pump).

As might be expected, the divergent-cone type has a performance characteristic more closely akin to an axial-flow pump, while the volute type has a performance more closely akin to a centrifugal pump.

Mixed-flow pumps are used when the head and flow required are intermediate between those suitable for axial-flow and centrifugal pumps.

5.5   *SPECIFIC SPEED*

The specific speed of a pump is defined mathematically as follows:

$$N_s = \frac{N\sqrt{\dot{V}}}{H^{\frac{3}{4}}}$$

Specific speed of a pump (5.1)

where　　N_s　=　specific speed
　　　　N　=　rotational speed of the impeller (rpm)
　　　　$\dot{V}$　=　volumetric flow rate (L/s)
　　　　H　=　head per stage developed by the pump (m)

Note　While units for specific speed are not usually quoted, specific speed is *not dimensionless* and its value depends on the units chosen for the various terms.

Specific speed is a useful parameter because it provides a single number which allows various types and sizes of rotodynamic pump to be compared. This is because *all geometrically similar rotodynamic pumps (regardless of their size) have the same specific speed at the point of operation where their efficiency is the same.*

This means that specific speed provides a useful comparison of the shape and design of a rotodynamic pump regardless of whether the pump is small or large. When selecting a pump for a particular application, calculation of the specific speed is the first step in determining whether a positive-displacement or rotodynamic pump would give the best efficiency. If a rotodynamic pump is chosen, specific speed provides a guide to the most suitable shape of impeller. The relationship between the type of pump and the specific speed is indicated in Figure 5.23.

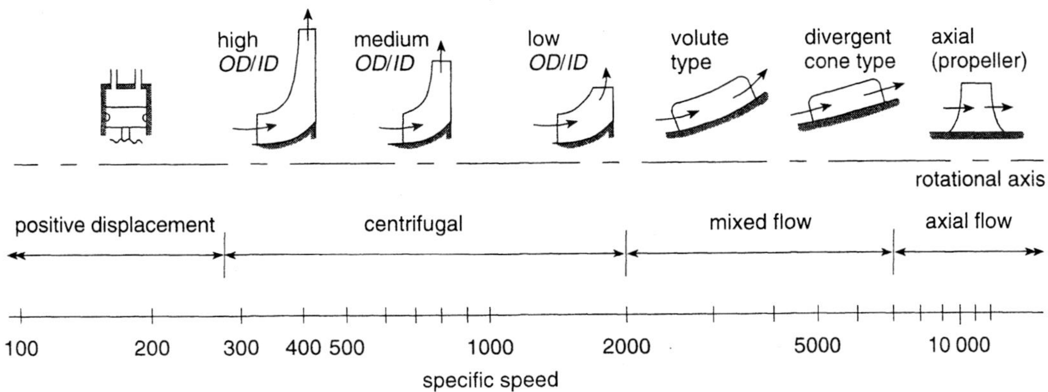

Fig. 5.23　*Specific speed of pumps*

Note that for low values of specific speed, positive-displacement pumps will give the best efficiency. There is some overlap between positive-displacement pumps and centrifugal pumps with impellers that have a high outlet/inlet diameter ratio (high *OD/ID* ratio). As the specific speed increases, the *OD/ID* ratio of centrifugal pumps reduces, until at a specific speed of about 2000, volute-type mixed-flow pumps are the most efficient. With increasing specific speed, divergent-cone types come into their own and at high values of specific speed (up to about 10000) axial flow pumps are the most efficient.

Notes

- The scale on the *x*-axis in Figure 5.23 is logarithmic (not linear).
- The dividing line between the various types of pumps should be regarded as fuzzy. For example, the lower limit of specific speed for a centrifugal pump is shown as 300. However, centrifugal pumps with specific speeds below 300 are commercially available but their efficiency is low. The essential point is that *the further the diversion from the optimum specific speed range, the lower the operating efficiency.*

- Specific speed applies to *geometrically similar pumps*, that is pumps in which all dimensions have been scaled up or down in the same ratio. If two pumps of different size are geometrically similar, all angles are the same and the pumps will have the same appearance. That is, if a photograph were taken of the smaller pump impeller, the larger geometrically similar impeller would be represented by an enlargement of the photograph.
- Geometric similarity requires that the surface finish (height of the roughness protuberances) be in the same proportions as the dimensions. In practice this is not usually the case because a certain method of manufacture (such as sand casting) tends to produce the same surface roughness, regardless of the size of the component. For this reason exact geometric similarity is seldom achieved with different pump sizes, and comparison between pumps of very different size using specific speed should not be regarded as exact.

Example 5.2

Determine a suitable pump for a system where the head is 50 m when the flow rate is 5 L/s. The pump will be directly driven by a four-pole electric motor rotating at 1450 rpm.

Solution

Calculate the specific speed:

$$N_s = \frac{N \sqrt{\dot{V}}}{H^{\frac{3}{4}}} = \frac{1450 \times \sqrt{5}}{50^{\frac{3}{4}}} = 172$$

Note from Figure 5.23 that this specific speed is below the minimum suitable for a rotodynamic pump. Even a centrifugal pump with a high *OD/ID* ratio will not give a reasonable efficiency with such a low specific speed. There are several possibilities as follows:

Option 1 Use a positive-displacement pump.

Option 2 Use a multistage centrifugal pump. In this case choose a specific speed of about 400 to give a suitable impeller profile and operating efficiency.

Now:

$$N_s = \frac{N \sqrt{\dot{V}}}{H^{\frac{3}{4}}}$$

$$\therefore 400 = \frac{1450 \times \sqrt{5}}{50^{\frac{3}{4}}}$$

$$\therefore H = 16.3 \text{ m}$$

Multistage centrifugal pumps are usually designed so that the head per stage is the same, hence the number of stages required is: $\dfrac{50}{16.3} = 3$. That is, a three-stage centrifugal pump could be used if a rotodynamic pump is preferred to a positive-displacement pump.

Option 3 Use three centrifugal pumps connected in series. This is an expensive option but achieves the same effect as one three-stage centrifugal pump.

Option 4 Use a 2-pole electric motor with a shaft speed of about 2900 rpm. In this case the specific speed will increase to 2 × 172 = 344 and this is within the range of reasonable efficiencies for a single-stage centrifugal pump, provided the impeller has a high *OD/ID* ratio.

The choice of options depends on factors such as the nature of the fluid, whether or not self-priming is required, pump and motor availability, cost, and so on.

Self-test problem 5.2

A pump is driven by a 4-pole electric motor rotating at 1450 rpm. Determine the most suitable pump for four different systems with head and flow rate requirements listed as (a)–(d) in the table:

System	Head (m)	Flow rate (L/s)
(a)	2	30
(b)	10	20
(c)	15	10
(d)	40	6

5.6 CAVITATION AND NET POSITIVE SUCTION HEAD

When the pressure of a liquid is gradually reduced, two effects can occur:

- Dissolved gases in the liquid break free as bubbles of gas escape from the liquid surface.
- The liquid vaporises, that is, changes phase from a liquid to a gas. With further reduction in pressure, a point can be reached where the vaporisation is rapid; in fact the liquid boils. The pressure at which this occurs is known as the saturation vapour pressure[4] (or simply vapour pressure or saturation pressure).

These two effects can occur on the suction side of a pumping system or within the pump itself when the fluid being pumped is a liquid. These effects tend to occur on the suction side (inlet) rather than the discharge side (outlet) because on the suction side the pressure is low and often below atmospheric, whereas on the discharge side the pressure is high and usually above atmospheric. However, as the liquid passes from the inlet to the outlet, vapour pockets formed in a low pressure region collapse again in a high pressure region.

The rapid formation and collapse of vapour pockets in a liquid is known as **cavitation**. When this occurs near a solid surface such as a pump impeller or casing, erosion and mechanical damage can occur. A cavitating pump makes a distinctive noise due to the vibration, and there is a marked drop in the efficiency and flow rate.

---

[4] The saturation pressure increases with temperature. That is, the higher the temperature of the liquid, the lesser is the *reduction in pressure* needed for boiling to occur. For example, for water at 20°C the pressure has to be reduced to 2.34 kPa (abs.), or by about 99 kPa below atmospheric for boiling to occur. However, at 60°C boiling occurs when the pressure is reduced to 20 kPa (abs.), or by about 81 kPa below atmospheric.

Notes

- Gas bubbles in a liquid do not collapse as readily as vapour pockets and are often carried through the pump to the outlet pipe. This is often known as aeration. In some cases gas bubbles contribute to cavitation in a pump.
- High-speed propellers such as those used in outboard motors can also cavitate. However, it is possible to design the propeller so that the vapour or gas bubbles collapse downstream of the propeller and hence do not damage it.

     Needless to say, cavitation should be avoided. In the design of a pump the factors that contribute to cavitation are:

- Sudden changes in direction such as sharp corners, because separation and voids occur when the liquid cannot follow the sudden direction change.
- Sudden changes in velocity (acceleration) of the liquid, because vapour pockets can form due to the inertia of the liquid. This is a particular problem with high-speed reciprocating pumps.
- High liquid velocities through the pump, because the higher the velocity, the lower is the pressure (the Bernoulli equation).

Net positive suction head available (*NPSHA*)

The most common method of evaluating whether or not a pump will cavitate is by using the net positive suction head available (*NPSHA*). The *NPSHA* of a pump is defined as *the difference between the total head at the inlet to the pump and the vapour pressure head of the liquid being pumped*. Using the inlet to the pump as a datum, there is no potential head, but the fluid has pressure head and velocity head. Therefore:

$$NPSHA = \frac{p_i}{\rho g} + \frac{v^2}{2g} - \frac{p_v}{\rho g}$$

or:

$$\boxed{NPSHA = \frac{p_i - p_v}{\rho g} + \frac{v^2}{2g}}$$ Net positive suction head available (5.2)

where p_i = pressure at the inlet to the pump in Pa (absolute)

            p_v = saturation vapour pressure of the liquid in Pa (absolute)

            ρ = density of the liquid in kg/m^3

            v = velocity of the liquid in the inlet pipe in m/s.

Notes

- In many cases the velocity head is negligible. For example, if the pump is pumping water and the suction velocity is 1 m/s, the velocity head is only about 0.05 m. However, if the suction velocity is 4 m/s, the velocity head is about 0.8 m and is more significant.
- The suction head is positive because absolute pressures are used (not gauge pressures) in the calculation of *NPSHA*. Often, at the pump inlet, the pressure is below atmospheric and therefore at a negative gauge pressure (vacuum). By convention, to avoid negative terms, absolute pressures are used, so the suction head is positive.
- A table of saturation vapour pressures for water (in the range of temperatures 0–100°C) is given in Appendix 13.

Net positive suction head required *(NPSHR)*

Cavitation can be avoided if the suction head at the inlet of the pump is above the value at which cavitation will occur. This value of the suction head is called the net positive suction head required *(NPSHR)* which is defined as: *the minimum value of the net positive suction head at the inlet to the pump in order to avoid cavitation.* The most reliable way of ascertaining the *NPSHR* for a certain pump is to test the pump using a standardised procedure. Sometimes pump manufacturers specify *NPSHR* by applying design rules based on past experience. Most manufacturers usually include a margin of safety when specifying the *NPSHR* for their pumps.

Example 5.3

A pump is tested for cavitation using the test rig illustrated in Figure 5.24. Water at 20°C is pumped and the inlet pipe diameter is 100 mm. The valve in the inlet line is gradually closed until the pump starts to cavitate. At this point the flow rate is 24 L/s and the vacuum gauge reads 62 kPa of vacuum. Determine for the given flow rate:
(a) the *NPSHA*;
(b) the *NPSHR* to be specified if a 5% safety margin is allowed.

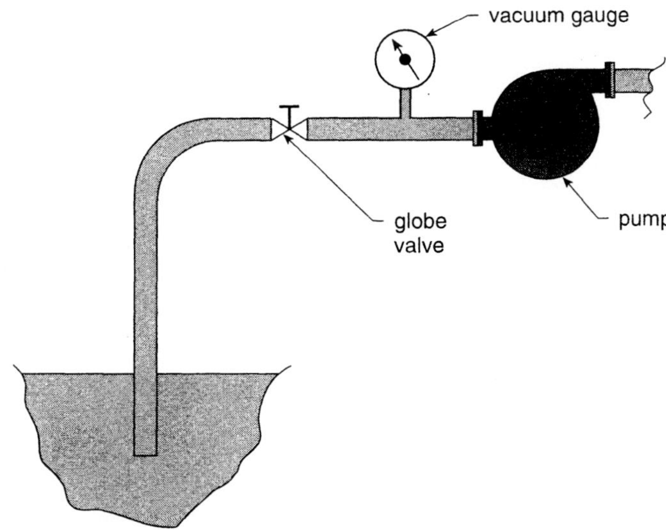

Fig. 5.24

Solution

(a) p_i = −62 kPa gauge = 39.3 kPa absolute
From Appendix 13, for water at 20°C, p_v = 2.34 kPa (absolute) and ρ = 998 kg/m³

$$v = \frac{0.024}{A(0.1)} = 3.06 \text{ m/s}$$

$$NPSHA = \frac{p_i - p_v}{\rho g} + \frac{v^2}{2g}$$

$$= \frac{39.3 \times 10^3 - 2.34 \times 10^3}{0.998 \times 10^3 \times 9.81} + \frac{3.06^2}{19.62}$$

$$= 3.775 + 0.476$$

$$= \mathbf{4.25 \text{ m}}$$

(b) If a 5% safety margin is to be allowed:

$$NPSHR = 4.25 \times 1.05 = \mathbf{4.46\ m}$$

Note   In this case, the velocity head is about 10% of the total *NPSHA*.

5.7 *PUMP PERFORMANCE — BASIC CONCEPTS*

Before looking in greater detail at the performance of positive-displacement and rotodynamic pumps, it is necessary to define the basic concepts used in the measurement of this performance.

Flow rate $\dot{V}$:   Flow rate (or discharge) is the amount of fluid pumped per unit time. Conventionally this is stated as a volumetric flow rate. Base SI units are m^3/s, but except for very large pumps, the flow rate is usually stated in litres per second (L/s). The volumetric flow rate can be converted to a mass flow rate (kg/s) by multiplying by the density of the fluid.

Head H: The head of a pump is the difference in total head of the fluid in metres (m) between the inlet (suction) and outlet (discharge) sides of the pump. Denoting the total head at inlet as H_1 and the total head at outlet as H_2, then $H = H_2 - H_1$. For a pump, H is positive because $H_2 > H_1$. As has been shown in Example 5.1, most of the head imparted to the fluid by a pump is in the form of pressure head due to the difference in pressure between the outlet and inlet sides of the pump. However, there may also be a small component of the total head due to a velocity head difference, or a potential head difference, or both.

Speed N: If the pump is driven by a rotating shaft, the speed of the pump is the speed of this shaft. The speed is usually expressed in revolutions per minute (rpm). Many pumps are directly coupled to electric motors and hence the most common pump speeds are about 2950 rpm (2 pole), 1450 rpm (4 pole), or 950 rpm (6 pole).[5]

Power P: The power of a pump is the input shaft power required to drive the pump. Base SI units are watts (W) but usually power is stated in kW. Shaft power is related to shaft torque by the basic mechanical power equation:

$$P = T\omega$$

where P = shaft power (W), T = shaft torque (Nm), and

$$\omega = \text{angular speed (rad/s)} = \frac{\pi N}{30}$$

Note   Prime movers such as electric motors or internal combustion engines are always rated on their shaft power output, and because they are not 100% efficient, their input energy rate is greater than their shaft power. For example, if a 10 kW electric motor has an efficiency of 75%, the energy input rate to the motor will be 10/0.75 = 13.3 kW.

Fluid power P_f: When referring to the power of a pump, power (unqualified) always means input power. However, as the fluid passes through the pump there is a change in

---

5 The synchronous speed (without slip) is 3000 rpm (2 pole), 1500 rpm (4 pole), and 1000 rpm (6 pole). The amount of slip varies with the load, so a precise motor speed cannot be stated.

fluid power due to the transfer of energy from the pump to the fluid. This change in fluid power is defined by the basic fluid power equation:

$$P_f = \dot{m}gH$$

where P_f = fluid power in W
$\dot{m}$ = mass flow rate of fluid in kg/s
H = pump head in m as defined above

If all of the head developed by the pump is in the form of pressure head, the fluid power equation can be written:

$$P_f = p\dot{V}$$

where p = pressure change to the fluid between inlet and outlet in Pa (either gauge or absolute)
$\dot{V}$ = volumetric flow rate in m^3/s

Efficiency η: The efficiency of a pump (unqualified) is the overall efficiency, that is, the ratio of fluid power to shaft power. It is given by:

$$\boxed{\eta = \frac{P_f}{P} = \frac{\dot{m}gH}{P}}$$ Pump efficiency (5.3)

The overall efficiency is the product of several other efficiencies:

- *Mechanical efficiency* η_m: Due to the fact that there is mechanical friction in pump bearings, seals and glands, some energy is dissipated in mechanical friction. For example, if 5% of the input power is dissipated in mechanical friction losses, then $\eta_m = 95\%$.
- *Hydraulic efficiency* η_h: As well as mechanical friction losses there are fluid friction losses. Their extent is measured by the hydraulic efficiency. For example, if 15% of the input power is dissipated in fluid friction, then $\eta_h = 85\%$.
- *Volumetric efficiency* η_v: The volume of fluid delivered by the pump is usually less than the volume of fluid displaced by the pump. This loss is measured by the volumetric efficiency. For example, if the theoretical flow rate through a positive-displacement pump is 5 L/s but the actual flow rate of fluid through the pump is 4.5 L/s, then $\eta_v = 90\%$.

With the figures given above, $\eta = \eta_m\,\eta_h\,\eta_v = 0.95 \times 0.85 \times 0.9 = 0.727 = 72.7\%$.

Self-test problem 5.3

A pump delivers 25 L/s of water with a head of 35 m and runs at a speed of 2950 rpm for 8 hours per day. The shaft power is 11.5 kW. The electric motor driving the pump has an efficiency of 84% and the cost of electricity is 11 cents per kWh. Determine:
(a) the shaft torque;
(b) the pump efficiency;
(c) the cost of electricity per day to run the pump.

5.8 *PERFORMANCE OF POSITIVE-DISPLACEMENT PUMPS*

It is customary to represent the performance of a positive-displacement pump graphically as lines drawn with outlet pressure (as gauge pressure) on the *x*-axis. Typical performance curves of this type are shown in Figure 5.25 (and also as Appendix 10).

These curves have been drawn for a gear pump of the external type pumping light oil and directly driven by a four-pole electric motor at a constant speed of 1450 rpm.

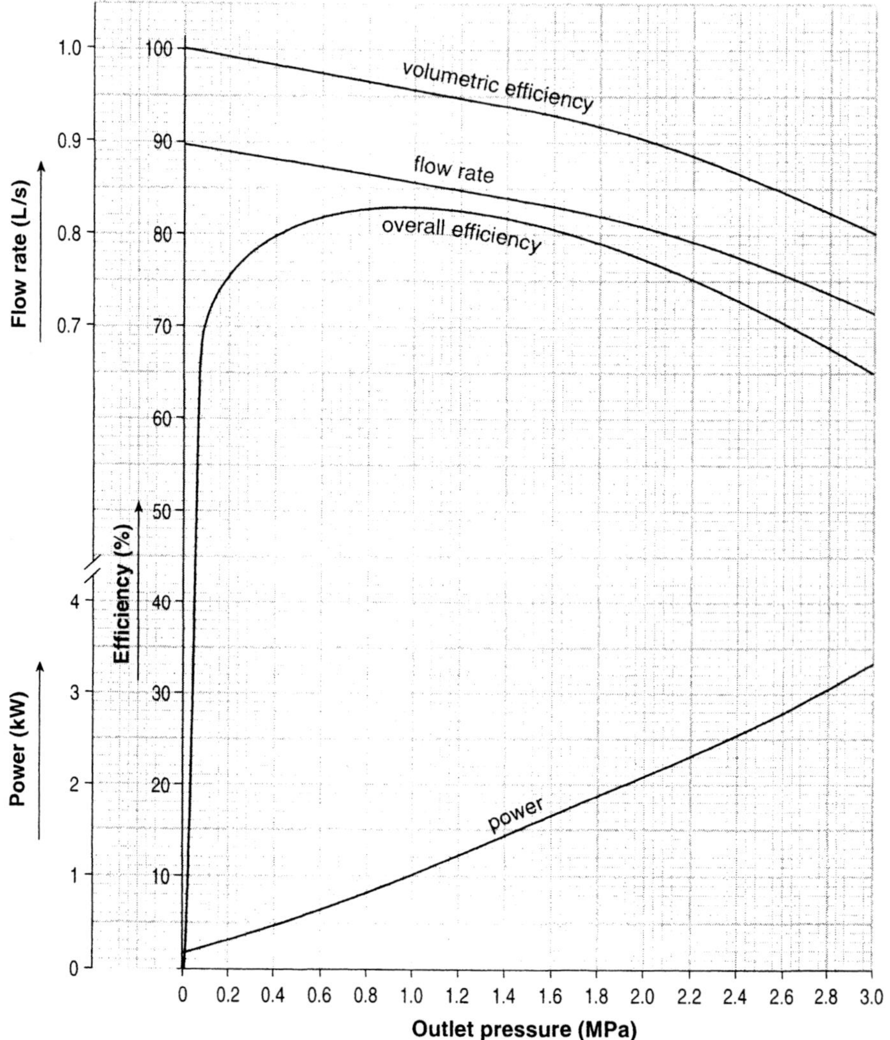

Fig. 5.25 *Performance curves for a gear pump running at constant speed (1450 rpm)*

Volumetric efficiency: If there was no internal leakage (or backflow) the volumetric efficiency would not change with outlet pressure. Then the volumetric efficiency would be 100% for the full range of outlet pressures. However, because of internal leakage the volumetric efficiency falls as the outlet pressure increases, and for this pump at maximum outlet pressure of 3 MPa the volumetric efficiency is 80%. With new pumps, higher volumetric efficiencies can be expected, but as wear occurs with usage, volumetric efficiency gradually decreases.

Volumetric flow rate: If there was no internal leakage the volumetric flow rate would not change with outlet pressure. However, because of internal leakage the curve for volumetric flow rate has the same shape as the curve for volumetric efficiency.

Power: As is typical with most positive-displacement pumps, the input power increases almost directly with the outlet pressure. That is, the power curve is almost a straight line.

Overall efficiency: At zero discharge pressure there is no fluid power because the pump is developing no head. However, input power is still required to overcome internal friction. Hence at zero outlet pressure the efficiency is zero. As the outlet pressure increases, the efficiency rapidly rises almost linearly to about 70%, after which the efficiency continues to rise, but at a slower rate, reaching a maximum value of about 83% at an outlet pressure of 800 kPa. After this point the efficiency falls due to the increasing losses.

Change in performance with speed

The performance curves in Figure 5.25 were drawn for a constant pump speed (1450 rpm). It is often necessary to change the pump speed in order to change the performance, and the effect of changes to the pump speed is shown in Figure 5.26.

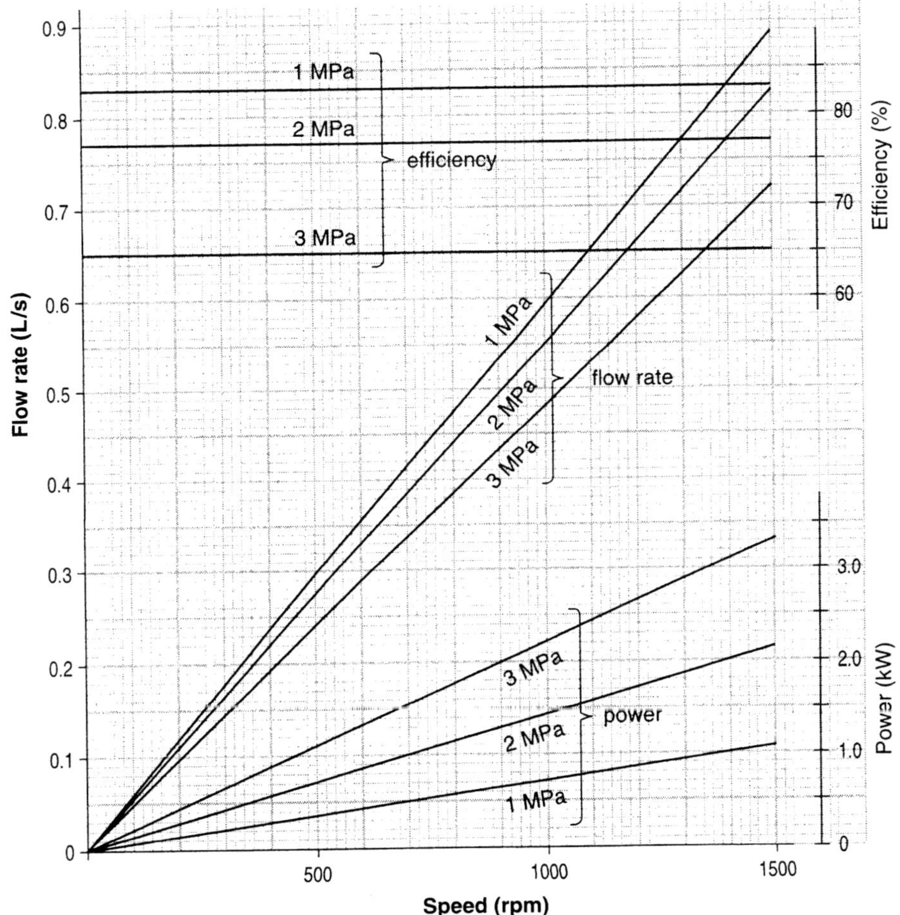

Fig. 5.26 *Performance of a gear pump at varying speeds*

Note that the curves approximate a series of straight lines at each outlet pressure value. The flow rate and power vary almost directly with speed. However, the efficiency remains approximately constant over the speed range for each value of the outlet pressure. This is an important advantage of positive-displacement pumps because for a given outlet pressure, the efficiency changes little over a wide range of speeds and flow rates.

Example 5.4

An external gear pump has gear rotors of outside diameter 50 mm, inside diameter 30 mm and width 40 mm. It is driven at 950 rpm and is pumping oil of relative density 0.9 which enters the pump at atmospheric pressure. When the outlet pressure is 1.5 MPa (gauge), the volumetric efficiency is 94% and the input power is 1.5 kW. Determine:

(a) the volumetric flow rate;
(b) the fluid power;
(c) the overall efficiency.

Solution

(a) For an external gear pump the volumetric displacement of *both* gears for one complete rotation is equal to the volume difference between the outside and inside diameters of *one* gear. This is because the space occupied by the teeth on one gear is equal to the space between the teeth on the other gear. Hence the volumetric displacement for one revolution of the gears is:

$$V = [A(50) - A(30)] \times 40 = 50.265 \times 10^3 \text{ mm}^3 = 50.265 \times 10^{-3} \text{ L}$$

Therefore, without leakage losses, $\dot{V} = 50.265 \times 10^{-3} \times \dfrac{950}{60} = 0.796$ L/s

Because $\eta_v = 94\%$, the actual flow rate $= 0.796 \times 0.94 =$ **0.748 L/s**

(b) Because all the fluid power is in the form of pressure head:

$$P_f = p\dot{V} = 1.5 \times 10^6 \times 0.748 \times 10^{-3} \text{ W} = \textbf{1.122 kW}$$

(c) $\eta_o = \dfrac{P_f}{P} = \dfrac{1.122}{1.5} = 0.748 = \textbf{74.8\%}$

 ### Self-test problem 5.4

A screw pump delivers 0.6 L/s of oil, *RD* 0.88, when fitted with a four-pole electric motor running at 1440 rpm. The outlet pressure is 500 kPa (g), the inlet pressure is −30 kPa (vac.) and the input power is 0.4 kW. Determine the efficiency.
If the pump is now powered by a two-pole electric motor running at 2950 rpm, determine:
(a) the new flow rate, and
(b) the new input power.
Assume the same inlet and outlet pressures and the same efficiency.

5.9 *PERFORMANCE OF ROTODYNAMIC PUMPS*

It is customary to show the performance of rotodynamic pumps graphically as curves drawn with flow rate on the *x*-axis. Typical performance curves of this type are shown in Figure 5.27 (and also as Appendix 11).

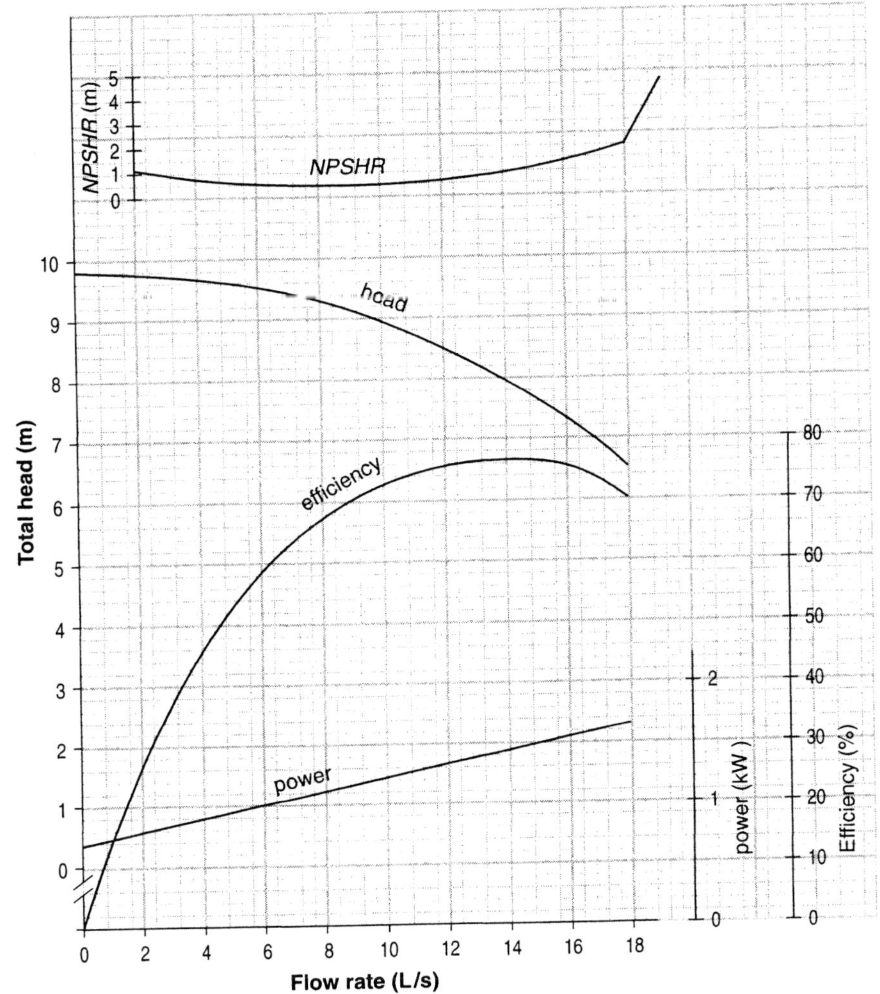

Fig. 5.27 *Performance of a centrifugal pump at constant speed (1450 rpm)*

In this case a centrifugal pump is pumping water[6] and is directly driven by a four-pole electric motor running at a constant speed of 1450 rpm.

---

[6] Most centrifugal pump performance data are for water as the fluid being pumped. If another fluid is being pumped, it is necessary to obtain performance curves relating to that fluid. If such curves are not available, it is possible to recalculate performance knowing the properties of the fluid, and thus draw another set of performance curves. The method of doing this is beyond the scope of this book.

Head: If there were no losses (ideal) the head–flow rate relationship would be a straight line whose slope depends on the impeller blade exit angle as shown in Figure 5.28.

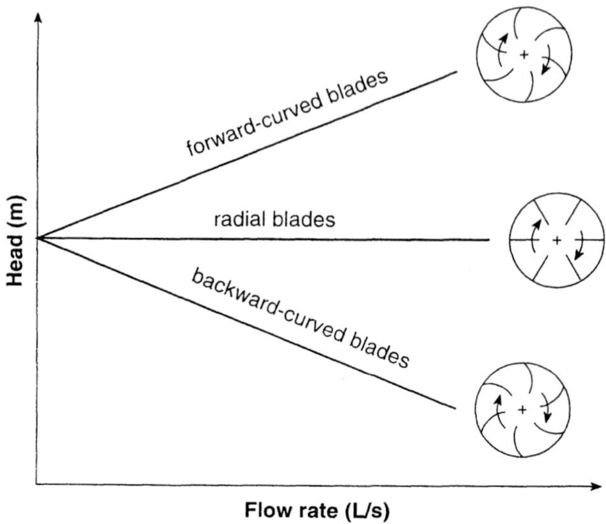

Fig. 5.28 *Ideal head–flow rate relationship for a centrifugal pump*

Because of losses. the actual head is less than the theoretical head, due to two factors:

- friction and circulatory flow losses as the fluid flows through the impeller;
- shock losses at entrance which arise because the fluid does not enter the impeller smoothly.

At design flow rate. the shock loss is virtually zero because the blade angles are set at the correct angle to avoid shock. At higher or lower flow rates than this, the radial velocity is different and because the impeller has blades set at a fixed angle, shock occurs at these flow rates. The meaning of shock-free entry is illustrated in Figure 5.29.

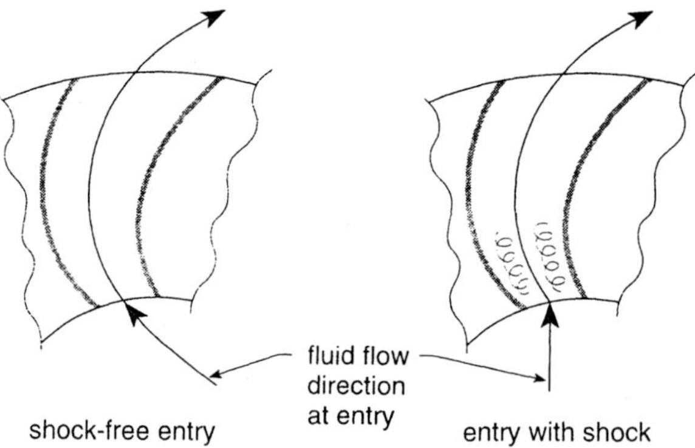

Fig. 5.29 *Shock-free entry and entry with shock*

The effect of these losses is to modify the ideal head–flow curve so that for a backward-bladed impeller it is as shown in Figure 5.30.

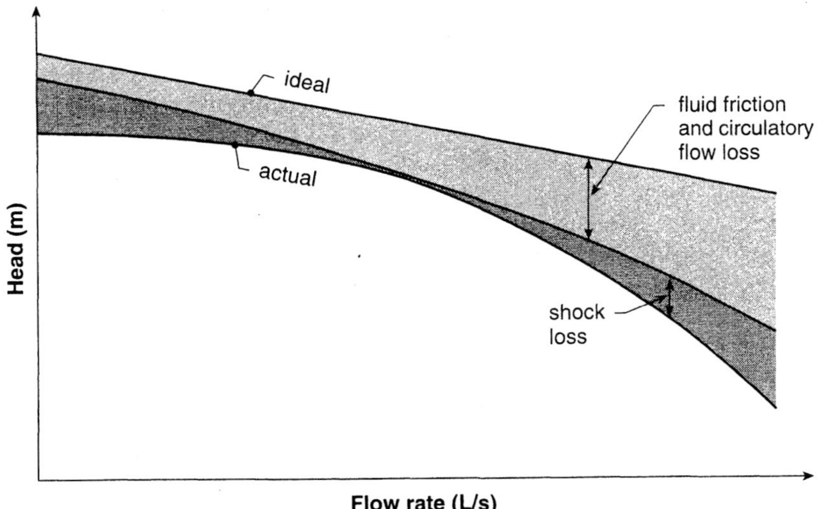

Fig. 5.30   *Ideal and actual head–flow relationship for a centrifugal pump (backward-bladed impeller)*

Note   The backward-bladed impeller invariably gives a continually falling head (falling characteristic). With radial or forward-inclined blades there may be a rising or a rising–falling characteristic.

Power: The power curve shows a continual increase in input power as the flow rate increases. There is usually a slight S-shape but on the scale used in Figure 5.27 the power curve appears almost linear.

Efficiency: At zero flow rate the efficiency is zero because there is no fluid power. As the flow rate increases, the efficiency also increases, reaching a maximum of about 76% at a flow rate of 14 L/s. As the flow rate increases above 14 L/s, the efficiency reduces further.

NPSHR: There is a small decline in the *NPSHR* as flow rate increases from zero, but then at a flow rate of about 10 L/s the *NPSHR* increases again. At higher flow rates the *NPSHR* rises more rapidly.

Example 5.5

Use the performance curves given in Appendix 11 to determine for a flow rate of 10 L/s:
(a) the head, power, efficiency and *NPSHR*;
(b) the fluid power, and using the input power reading from the performance curves, calculate the efficiency and compare it with the value read off the curves.

Solution

(a) At 10 L/s,
$$H = \textbf{8.95 m}$$
$$P = \textbf{1.2 kW}$$
$$\eta = \textbf{73\%}$$
$$NPSHR = \textbf{0.5 m}$$

(b) $P_f = mgH = 10 \times 9.81 \times 8.95 = \textbf{878 W}$

$\eta = \dfrac{P_f}{P} = \dfrac{878}{1200} = \textbf{73\%}$ (which is the same value as that read off the curves)

Effect of changing the impeller diameter

Manufacturers of centrifugal pumps frequently derive a number of different performance characteristics by machining the outside of the impeller to a smaller diameter. This achieves manufacturing economies because it is more cost effective to machine down one impeller casting to various sizes than to make each impeller as a separate casting. The remainder of the pump stays the same, so other sizes such as the inlet and outlet diameters, or volute dimensions, do not change. The performance of a centrifugal pump with various impeller diameters, pumping water and running at 1450 rpm, is shown in Figure 5.31 and also as Appendix 12.

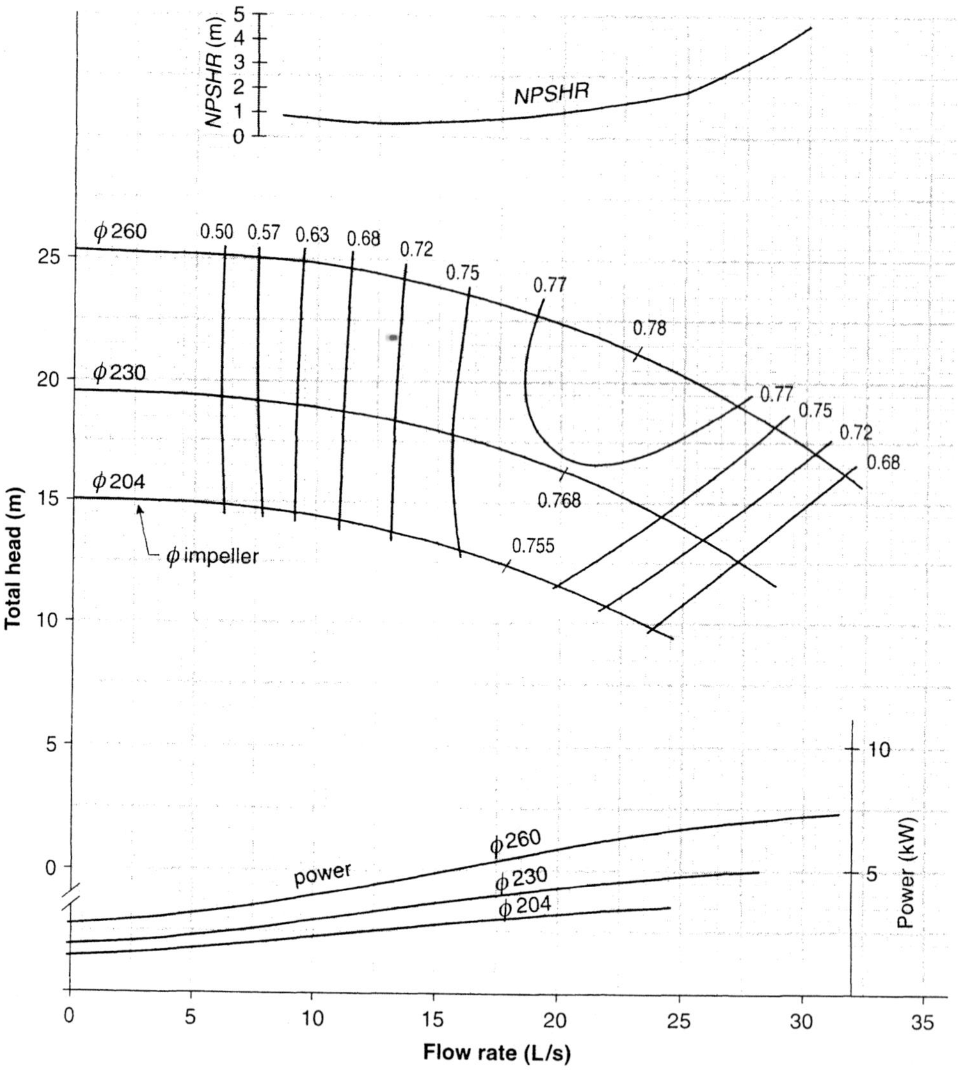

Fig. 5.31 *Performance of a centrifugal pump with various impeller diameters at constant speed (1450 rpm)*

For this particular pump, three impeller diameters are standard: 260 mm, 230 mm, and 204 mm. Note that there are three head curves and three power curves (one for each impeller size). However, there is only one *NPSHR* curve because *NPSHR* does not change significantly with impeller size. While three efficiency curves could be drawn, it is common practice to show the efficiency as a series of iso-efficiency lines as this avoids filling the chart with too many lines and makes it easier to read.

 ## *Self-test problem 5.5*

Use the performance curves given in Appendix 12 for each impeller size when the flow rate is 20 L/s to determine:
(a) the head, power, efficiency and *NPSHR*;
(b) the fluid power, and then, using the input power readings from the performance curves, calculate the efficiency and compare it with the value read off the curves.

Change in performance with speed

Manufacturers of rotodynamic pumps may produce a number of performance curves for different common pump speeds. However, when such curves are not available, or the speed is different to that given by the manufacturer, new performance characteristics can be obtained from available performance curves by using the so-called *affinity relations*. These apply when the efficiency and the specific speed are the same, that is:

$$\eta_1 = \eta_2$$
$$N_{s_1} = N_{s_2}$$

Then the affinity relations are:

$$\boxed{\frac{\dot{V}_2}{\dot{V}_1} = \frac{N_2}{N_1}}$$ Volume–speed relationship for a rotodynamic pump (5.4)

$$\boxed{\frac{H_2}{H_1} = \left[\frac{N_2}{N_1}\right]^2}$$ Head–speed relationship for a rotodynamic pump (5.5)

$$\boxed{\frac{P_2}{P_1} = \left[\frac{N_2}{N_1}\right]^3}$$ Power–speed relationship for a rotodynamic pump (5.6)

The affinity relations can be expressed in words as follows:
- The flow rate is directly proportional to the speed.
- The head is directly proportional to the square of the speed.
- The power is directly proportional to the cube of the speed.

Notes
- The affinity relations should not be regarded as exact, but rather a guide to the expected performance of the pump at the new speed.

- The affinity relations apply to fans and blowers as well as to pumps, provided that the gas density remains constant. The affinity relations are then usually known as the *fan laws*.
- The affinity relations do not apply to positive-displacement pumps (only rotodynamic ones).

Example 5.6

At the point of maximum efficiency, a centrifugal pump running at 1450 rpm delivers 38 L/s of water with a head of 20 m. The power input is 9.2 kW and the efficiency is 81%. Determine the flow rate, head and power at the maximum efficiency point if the pump speed is increased to 2900 rpm. Show that the efficiency and specific speed are the same.

Solution

$$\frac{\dot{V}_2}{\dot{V}_1} = \frac{N_2}{N_1}$$

$$\therefore \dot{V}_2 = 38 \times \frac{2900}{1450} = \textbf{76 L/s}$$

$$\frac{H_2}{H_1} = \left[\frac{N_2}{N_1}\right]^2$$

$$\therefore H_2 = 20 \times \left[\frac{2900}{1450}\right]^2 = \textbf{80 m}$$

$$\frac{P_2}{P_1} = \left[\frac{N_2}{N_1}\right]^3$$

$$\therefore P_2 = 9.2 \times \left[\frac{2900}{1450}\right]^3 = \textbf{73.6 kW}$$

$$\eta_2 = \frac{P_{f_2}}{P_2} = \frac{\dot{m}_2 g H_2}{P_2} = \frac{76 \times 9.81 \times 80}{73\,600} = 0.81 = \textbf{81\%}$$

Hence the efficiency has not changed

$$\text{Specific speed} \quad N_s = \frac{N\sqrt{\dot{V}}}{H^{\frac{3}{4}}}$$

$$N_{s_1} = \frac{1450 \times \sqrt{38}}{20^{\frac{3}{4}}} = \textbf{945}$$

$$N_{s_2} = \frac{2900 \times \sqrt{76}}{80^{\frac{3}{4}}} = \textbf{945}$$

Hence the specific speed has not changed

Self-test problem 5.6

The performance of a centrifugal pump pumping water and running at a speed of 1420 rpm is as follows:

Flow rate (L/s)	0	10	20	30
Head (m)	23	22	20.5	15.5
Power (kW)	3.8	4.3	5.7	6.5

(a) Draw the performance curves to scale for this pump, that is, plot head, power and efficiency against flow rate on the x-axis. Use full lines.
(b) Predict the performance of this pump if it were run at 950 rpm, that is, draw the performance curves for the pump at this speed. Draw these curves on the same sheet of graph paper as used for (a) and use broken lines.
(c) Calculate the specific speed for a flow rate of 20 L/s at 1420 rpm, and for the corresponding point of operation at 950 rpm, and show that they are the same.

5.10 *MOTORS AND TURBINES*

As already stated, the essential difference between a motor or turbine, and an actuator, pump or compressor, is that in the former case fluid power is converted to mechanical power, and in the latter case mechanical power is converted to fluid power. In some cases operation in either mode is possible, simply by reversing the flow of the fluid. For example, this is possible with a gear pump (positive-displacement) or a mixed-flow pump (rotodynamic). In other cases this is not possible, for example, with a screw pump (positive-displacement) or a Pelton wheel turbine (rotodynamic).

The performance terms and criteria for motors or turbines are the same as for pumps, except that the efficiency and specific speed are defined differently. For a motor or turbine these are defined by the following equations:

$$\eta = \frac{P}{P_f}$$

Motor or turbine efficiency (5.7)

$$N_s = \frac{N\sqrt{P}}{H^{\frac{5}{4}}}$$

Specific speed of a motor or turbine (5.8)

While most manufacturing or engineering applications use a variety of pumps, fluid motors and turbines are less common. This is because motive power is primarily provided by electric motors (or sometimes heat engines). However, linear or rotary actuators are often used to provide the force needed for many manufacturing and engineering applications such as presses, moulding machines, clamping devices, and so on.

Summary

Fluid flows naturally from a higher pressure or elevation to a lower one. Useful energy is lost when this occurs because energy is dissipated in turbulence and fluid friction. Conversion of this energy into useful mechanical energy requires a motor, turbine or actuator. In order to reverse the free-flow direction, that is, to cause a fluid to flow from a lower elevation or pressure to a higher one, requires a pump, fan or compressor. Fluid machinery can thus be divided into two broad classes:

Machines that convert mechanical power into fluid power

- Positive-displacement pumps, such as piston pumps, plunger pumps, diaphragm pumps, gear pumps, lobe pumps, geroter pumps, vane pumps, screw pumps and peristaltic pumps.
- Rotodynamic pumps, such as axial-flow pumps, centrifugal or radial-flow pumps, or mixed-flow pumps.
- Compressors, which are of the same general types as pumps, but are used with gases, to increase the pressure.
- Fans and blowers, which are usually of the lobe or rotodynamic type and are used to move large volumes of gases with little pressure increase.
- Propellers, which are essentially axial-flow machines designed to obtain a large thrust.

Machines that convert fluid power to mechanical power

- Fluid motors, which are essentially positive-displacement pumps (usually of the piston or gear type) used in reverse.
- Turbines, which are essentially rotodynamic pumps used in reverse.
- Actuators (pneumatic or hydraulic cylinders) that usually provide linear reciprocating motion but sometimes provide reciprocating rotary motion.

Selection of the appropriate fluid machine in any practical application involves consideration of a large number of factors, and balancing their relative advantages and disadvantages to obtain an optimum solution over the life of the equipment.

Also, specific speed provides a useful guide to the most appropriate fluid machine for a given application. Specific speed is defined by Formulas 5.2 and 5.8:

$$N_s = \frac{N \backslash \dot{V}}{H^{\frac{3}{4}}} \quad \text{pump (5.2)}; \quad N_s = \frac{N \backslash \bar{P}}{H^{\frac{5}{4}}} \quad \text{motor or turbine (5.8)}$$

The relationship between specific speed and the type of fluid machine has been established by practical observation and is usually presented in diagrammatic form as in Figure 5.23 for pumps. While fluid machinery will still function when outside the appropriate specific speed range, the efficiency will not be as high as when the machinery is operating within the appropriate specific speed range.

A problem that can occur with pumps of both the positive-displacement and rotodynamic types is cavitation, where pockets of vapour, formed in a low pressure region, collapse in a high pressure region. This not only lowers the operating efficiency, but also can cause noise, vibration and mechanical damage. The most common way to evaluate whether or not a pump is likely to cavitate is to calculate the net positive suction head available (*NPSHA*) as defined by Formula 5.2. Pump manufacturers establish the net positive suction head required (*NPSHR*) in order to prevent cavitation in the pump. The value of the *NPSHR* can vary with flow rate and is usually shown graphically on the performance curves for a pump. Provided that *NPSHA* > *NPSHR*, the pump should not cavitate under operating conditions.

The performance of a fluid machine is usually shown graphically (performance curves). These curves show the variation between the head (or pressure), flow rate, power, efficiency and *NPSHR*. In the case of centrifugal pumps, manufacturers often use one impeller casting which is then machined down in order to produce a number of different performance characteristics for the same basic pump. When this is done, the efficiency is usually shown as a series of iso-efficiency lines.

The same performance criteria apply whether the fluid machine is converting fluid power to mechanical power or vice versa. However, the main differences are the way in which the specific speed is calculated (compare Formulas 5.2 and 5.8) and the way in which efficiency is calculated. The efficiency formulas are inverted when calculating the efficiency of a motor or turbine rather than a pump, as follows:

$$\eta = \frac{P_f}{P} \quad \text{pump} \quad (5.3); \qquad \eta = \frac{P}{P_f} \quad \text{motor or turbine} \quad (5.7)$$

Frequently, a pump will be operated at a speed different to that applicable to the performance curves. In such cases the new performance can be estimated using the affinity relations that apply when the specific speed and efficiency are the same. These relations are expressed mathematically by Formulas 5.4, 5.5 and 5.6, or, in words: *flow rate is proportional to speed, head is proportional to the speed squared, and power is proportional to the speed cubed.*

 Problems

5.1 **(a)** Briefly explain the difference between positive-displacement and rotodynamic fluid machinery.
   (b) For which applications is each type most suitable?
   (c) Briefly explain the difference between parallel and series arrangements.
   (d) When is each arrangement most suitable?

5.2 Outline at least ten factors that should be considered when deciding on the most suitable fluid machine for a given application.

5.3 **(a)** Define a pump.
   (b) What is the difference between a pump and a compressor?
   (c) In what form is most of the head increase imparted by a pump?
   (d) What influences the pressure developed by a pump when it is operating, and also the maximum pressure that can be developed?
   (e) Describe the conditions under which *all* of the head increase across a pump will be in the form of pressure head.

5.4 **(a)** Name eight types of positive-displacement pump.
   (b) Name three types of rotodynamic pump.
   (c) List at least five advantages and five disadvantages of positive-displacement pumps, compared with rotodynamic pumps.
   (d) What is an actuator, and what are the most common types of actuator?
   (e) What is the essential difference between a fluid motor and a turbine?

5.5 (a) With the aid of neat sketches, briefly describe the operation of a fixed-volume piston pump and a variable-volume piston pump.
(b) How is excessive outlet pressure prevented in each case?

5.6 (a) With the aid of a neat sketch, briefly describe the operation of a plunger pump.
(b) What is the essential difference between a plunger pump and a piston pump?

5.7 (a) With the aid of a neat sketch, briefly describe the operation of a diaphragm pump.
(b) What are the main advantages and disadvantages of a diaphragm pump compared to a piston pump?

5.8 With the aid of a neat sketch, briefly describe the difference between a single-acting and a double-acting piston pump.

5.9 With the aid of a neat sketch, briefly describe the operation of a gear pump of both the internal and external gear type.

5.10 (a) With the aid of a neat sketch, briefly describe the operation of a lobe pump.
(b) What are the advantages and disadvantages of a lobe pump, compared to a gear pump?

5.11 With the aid of a neat sketch, briefly describe the operation of a geroter pump.

5.12 (a) With the aid of a neat sketch, briefly describe the operation of a vane pump.
(b) Briefly describe the three variations of the vane pump.

5.13 (a) How many screws (or rotors) are used with screw pumps?
(b) What is the essential difference between screw pumps with timed and untimed rotors?
(c) What is the essential difference between a single- or double-ended screw pump?
(d) What are the main advantages and disadvantages of screw pumps, compared to the other types of positive-displacement pump?

5.14 (a) With the aid of a neat sketch, briefly describe the operation of a peristaltic pump.
(b) What are the main advantages and disadvantages of peristaltic pumps, compared to the other types of positive-displacement pump?
(c) What are the main applications of peristaltic pumps?

5.15 (a) With the aid of a neat sketch, briefly describe the essential difference between axial-flow, radial-flow and mixed-flow rotodynamic pumps.
(b) For what applications is each of these types most suitable?

5.16 (a) With the aid of a neat sketch, briefly describe the operation of a centrifugal pump. On your sketch, show the path of the fluid relative to the impeller, and relative to the casing.
(b) With the aid of a neat sketch, show the three blade angles possible with centrifugal pumps.
(c) For each of the blade angles shown in (b), sketch the ideal head–flow rate relationship.

5.17 (a) With the aid of neat sketches, explain the difference between the following centrifugal pump impellers: open, closed (shrouded), single suction, and double suction.
(b) Sketch the ideal head–flow rate relationship for a centrifugal pump with backward blades, and show how the actual relationship can be derived by the inclusion of losses.

5.18 **(a)** Draw a neat sketch of a two-stage centrifugal pump and a two-stage axial flow pump.
(b) What benefits arise from staging?
(c) What is the advantage of using a multiple-stage pump rather than a number of single-stage pumps connected in series?
(d) What is meant by self-priming?
(e) Name the group of pumps that will self-prime, and the group that will not.

5.19 **(a)** Briefly describe the phenomenon of cavitation in a pump.
(b) Define net positive suction head available for a pump.
(c) Define net positive suction head required for a pump.
(d) What effect does fluid temperature have on the tendency for a pump to cavitate?
(e) What undesirable effects occur as a result of cavitation?

5.20 **(a)** What is meant by shock-free entry? Illustrate your answer by drawing a diagram showing shock-free entry, and entry with shock.
(b) State the affinity relations in words.
(c) What is necessary for the affinity relations to be valid for gases?

5.21 A pump delivers 50 L/s of water with a head of 55 m and runs at a speed of 1470 rpm for 8 hours per day. The efficiency is 78%. The electric motor driving the pump has an efficiency of 88% and the cost of electricity is 10.5 cents per kWh. Determine:
(a) the shaft power;
(b) the shaft torque;
(c) the cost of electricity per day to run the pump.
(a) 34.6 kW (b) 225 Nm (c) $33.00

5.22 In a single-stage, single-acting piston pump, the piston has a diameter of 100 mm and a stroke of 25 mm. The crankshaft rotates at 720 rpm.
(a) Determine the theoretical flow rate of liquid through the pump in L/s.
(b) If the velocity of the liquid in the inlet and outlet ports is not to exceed an average value of 3 m/s over the complete cycle, determine the minimum internal diameter of these ports.
(a) 2.36 L/s, (b) 31.6 mm

5.23 A vane pump has a displacement of 0.08 L per revolution of the shaft. At a speed of 1440 rpm it delivers 1.75 L/s of liquid. The pressure increase across the pump is 4 MPa and the pump efficiency is 82%.
Determine:
(a) the volumetric efficiency;
(b) the input power;
(c) the mechanical efficiency if the hydraulic efficiency is 95%.
(a) 91.15% (b) 8.54 kW (c) 94.7%

5.24 An external-gear pump has gear rotors of outside diameter 45 mm, inside diameter 25 mm, and width 30 mm. It is driven at 1250 rpm and is pumping oil of relative density 0.9, which enters the pump at atmospheric pressure. When the pressure increase across the pump is 1.2 MPa, the volumetric efficiency is 95% and the overall efficiency is 76%.
Determine:
(a) the volumetric flow rate;
(b) the fluid power;
(c) the shaft power.
(a) 0.653 L/s (b) 783 W (c) 1.03 kW

5.25 If the pump in Problem 5.24 is now run as a motor with a pressure decrease of 1.2 MPa across the motor, calculate the shaft power, assuming all other factors remain the same.
595 W

5.26 A screw pump delivers 0.5 L/s of oil, *RD* 0.92 when fitted with a six-pole electric motor running at 940 rpm. The outlet pressure is 700 kPa (g), the inlet pressure is −20 kPa (vac.) and the efficiency is 82%. If the pump is now powered by a four-pole electric motor running at 1430 rpm, determine the new flow rate and input power. Assume the same inlet and outlet pressures and the same efficiency.
0.761 L/s, 668 W

5.27 A hydraulic motor operating at a speed of 1750 rpm has a volumetric efficiency of 90% when the pressure drop across the motor is 7 MPa. The actual flow rate of oil through the motor is 0.26 L/s and the actual torque is 7.5 Nm. Determine:
(a) the power output;
(b) the overall efficiency;
(c) the swept volume per revolution.
(a) 1.37 kW (b) 75.5% (c) 9.9 mL

5.28 A double-acting pneumatic cylinder attached to a casting machine has a bore of 80 mm, a stroke of 325 mm, and a rod diameter of 20 mm. It is supplied with air at a pressure of 120 kPa (g). Frictional losses are 5%. Each stroke takes a time of 5 s and there is a dwell time of 10 s between strokes. Determine:
(a) the force exerted by the rod on the forward stroke;
(b) the force exerted by the rod on the return stroke;
(c) the power output on the forward stroke;
(d) the theoretical volume of compressed air required to operate the machine continuously for one hour.
(a) 573 N (b) 537 N (c) 37.2 W (d) 0.38 m^3

5.29 Determine the specific speed of the following centrifugal pumps:
(a) A single-stage pump running at 1450 rpm, with a total head of 12.8 m and a flow rate of 15 L/s.
(b) A two-stage pump running at 1450 rpm, with a total head of 24 m and a flow rate of 25 m^3/h.
(a) 830 (b) 593

5.30 A centrifugal pump is to deliver 9 L/s with a head of 40 m at a shaft speed of 960 rpm. If a specific speed of about 400 is required, how many stages are necessary?
Three

5.31 The centrifugal pump whose performance curves are given in Appendix 12 has an inlet diameter of 125 mm and an outlet diameter of 80 mm. Calculate:
(a) the inlet and outlet velocities when the flow rate is 10 L/s;
(b) the inlet and outlet velocities when the flow rate is 30 L/s;
(c) the velocity head as a percentage of the total head for the 260 mm diameter impeller at a flow rate of 30 L/s.
(a) 0.815 m/s, 1.99 m/s (b) 2.44 m/s, 5.97 m/s (c) 8.76%

5.32 For the centrifugal pump whose performance curves are given in Appendix 12, calculate the specific speed at the maximum efficiency point for each impeller size. Comment on the results with reference to Figure 5.23.
260 mm: 709; 230 mm: 811; 204 mm: 912

5.33 For the centrifugal pump whose performance curves are given in Appendix 12, show the efficiency for the 204 mm and 260 mm impellers as curves with efficiency on the *y*-axis and flow rate on the *x*-axis. Draw the two curves on the same sheet of graph paper and label them.

5.34 On a cavitation test for a pump, water at 24°C flows through the inlet pipe with velocity 3.5 m/s and pressure 57 kPa below atmospheric when cavitation commences. Determine:
(a) the *NPSHA*;
(b) the *NPSHR* if a 5% safety margin is to be provided.
Use Appendix 13 for the saturation vapour pressure and density of water at the given temperature.
(a) 4.85 m (b) 5.09 m

5.35 When water at 10°C flowed through a pump, the pump started to cavitate when a gauge at the inlet to the pump showed a pressure of 45 kPa below atmospheric. If the same pump were pumping water at a temperature of 60°C at the same rate, what would the pressure gauge now show at the onset of cavitation?
Use Appendix 13 for the saturation vapour pressure and density of water at the given temperatures.
−27.2 kPa

5.36 For the pump illustrated in Figure P5.36, the flow rate is 4.5 L/s of sea water (*RD* 1.04). The pressure at the pump inlet is −30 kPa (g) and at the pump outlet is 740 kPa (g). Determine the following pumping heads:
(a) pressure head;
(b) velocity head;
(c) potential head;
(d) total head.

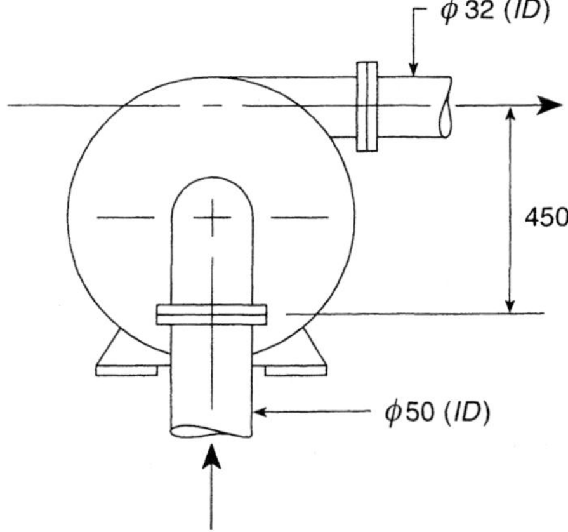

Fig. P5.36

(a) 75.5 m (b) 1.33 m (c) 0.45 m (d) 77.3 m

5.37 The pump in Problem 5.36 is a centrifugal type and the speed is 2930 rpm.
(a) Calculate the specific speed.
(b) Calculate the efficiency if the input power is 9 kW.
(c) Explain why you think the efficiency is so low.
(a) 238 (b) 39.4%

5.38 The performance of a centrifugal pump pumping water, and running at a speed of 1465 rpm, is:

Flow rate (L/s)	0	5	10	15	20	25	30
Head (m)	24.1	24.0	23.9	23.3	22.3	20.8	18.7
Power (kW)	3.0	3.6	4.4	5.3	6.2	7.0	7.6

(a) Draw the performance curves to scale for this pump, that is, plot head, power and efficiency against flow rate on the x-axis. Use full lines.

(b) Predict the performance of this pump if it were run at 1800 rpm, that is, draw the performance curves for the pump at this speed. Draw these curves on the same sheet of graph paper as used in (a) and use broken lines.

(c) Calculate the specific speed for a flow rate of 30 L/s at 1465 rpm and for the corresponding point of operation at 1800 rpm, and show that they are the same.

(a)

Flow rate (L/s)	0	5	10	15	20	25	30
Efficiency (%)	0	32.7	53.3	64.7	70.6	72.9	72.4

(b)

Original flow rate (L/s)	0	5	10	15	20	25	30
New flow rate (L/s)	0	6.14	12.3	18.4	24.6	30.7	36.9
New head (m)	36.4	36.2	36.1	35.2	33.7	31.4	28.2
New power (kW)	5.56	6.68	8.16	9.83	11.5	13.0	14.1
New efficiency (%)	0	32.7	53.3	64.7	70.6	72.9	72.4

Pumping systems

Objectives

On completion of this chapter you should be able to:

- select a suitable pump for a fluid pumping system;
- determine the duty point for a pumping system by combining the system head curve with the pump performance curve;
- determine the flow rate, head, power and efficiency at the duty point;
- calculate the energy cost of pumping, and choose a pump that will give the lowest energy cost for a certain pumping duty;
- determine the effect on the flow rate and other parameters in the system when a flow control valve setting is changed, and redraw the system curve to ascertain the new duty point;
- calculate the available net positive suction head (*NPSHA*) and hence determine the likelihood of cavitation;
- describe the suction line conditions that will minimise the likelihood of cavitation.

Introduction

In Chapter 5 the various types of positive-displacement and rotodynamic pumps were described and their important advantages and disadvantages outlined. In Chapter 2 the method of calculating the system head at a given flow rate through the system was described. It was shown that the system head can be represented by an equation (the system head equation). This enables the system head to be plotted graphically as a function of the flow rate through the system.

This chapter deals with selection of a suitable pump for a given system and how the system head curve and pump performance curve can be combined to determine the operating point (duty point). Before doing so, the derivation of the system head equation will be revised.

6.1 *SYSTEM HEAD EQUATION*

The system head is the sum of two components:

- A static component H_{stat} consisting of the pressure head plus the potential head. This component does not change with flow rate (or velocity).
- A dynamic component H_{dyn} consisting of the head loss plus the change in velocity head (if this occurs). If the friction factor is constant, this component varies as the square of the flow rate (and velocity).

The system head can be expressed mathematically by Formulas 2.8 and 2.9:

$$H = A + Bv^2 \qquad \text{System head equation—in terms of velocity (2.8)}$$

$$H = A + C\dot{V}^2 \qquad \text{System head equation—in terms of flow rate (2.9)}$$

In these equations:

A is a constant and equal to the static head H_{stat};

B and C are factors that will have a constant value if the friction factor does not vary with flow rate.

Once the system head equation has been written, the system head curve can be graphed to scale to show the relationship between system head on the y-axis and flow rate (or velocity) on the x-axis.

In the system head curve, constant A is the intercept on the y-axis and represents the system head when the velocity and flow rate are zero. Constant A will usually have a positive value, and when this is so, the curve starts from a position above the origin on the y-axis. However, constant A can be zero or negative, in which case the curve will start from the origin, or below the origin on the y-axis (respectively).

Factors B and C determine the steepness or slope of the curve, and the larger their values the more steeply the curve will rise. If the friction factor is constant over the range of plotted values (lies in the constant zone in the Moody diagram) the system head curve will be exactly parabolic.

The method of obtaining the system head equation and drawing the system head curve is revised in Example 6.1.

Example 6.1

For the system shown in Figure 6.1 (opposite), water at 40°C is pumped from a storage tank through a spray head located in a pressurised vessel.

(a) Determine the system head equation in the form: $H = A + C\dot{V}^2$.

(b) Draw the system head curve for flow rates in the range 0–30 L/s.

Data

Piping: galvanised steel, diameter 125 mm, length 5 m on the suction side and 65 m on the discharge side.

Spray head K factor: 15 (includes allowance for loss of velocity head in the spray).

Assumption

Constant friction factor based on a flow rate of 15 L/s.

Solution

(a) From Appendix 4:

$$\varepsilon = 0.15 \text{ mm}, \therefore \varepsilon_R = \frac{0.15}{125} = 1.2 \times 10^{-3}$$

$$\text{When } \dot{V} = 15 \text{ L/s}, v = \frac{0.015}{A(0.125)} = 1.222 \text{ m/s}$$

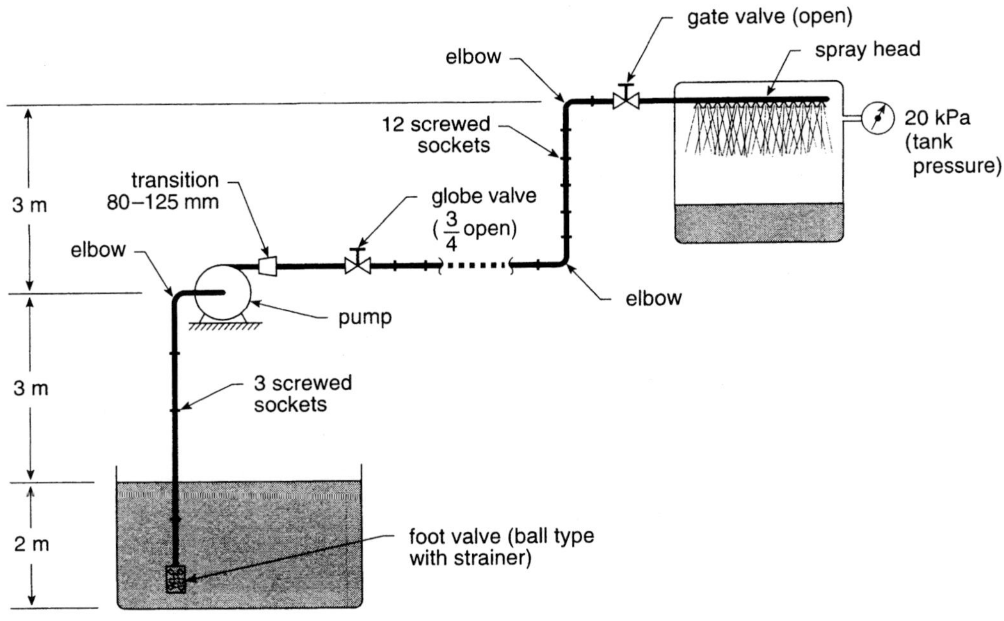

Fig. 6.1

From Appendix 13, for water at 40°C: $\mu = 0.66 \times 10^{-3}$ Pas, $\rho = 992$ kg/m^3

$$Re = \frac{vd\rho}{\mu} = \frac{1.222 \times 0.125 \times 992}{0.66 \times 10^{-3}} = 0.23 \times 10^6$$

$$f = 0.0055\left[1 + \left(20\,000\varepsilon_R + \frac{10^6}{Re}\right)^{\frac{1}{3}}\right]$$

$$= 0.0055\left[1 + \left(20 \times 1.2 + \frac{1}{0.23}\right)^{\frac{1}{3}}\right]$$

$$= 0.0223$$

Using Appendix 6, the following table can be drawn up:

Fitting	K factor	Number off	ΣK
Spray head	15	1	15
Ball foot valve with strainer	7	1	7
Screwed socket	0.03	15	0.45
Globe valve ($\frac{3}{4}$ open)	8	1	8
Gate valve (open)	0.2	1	0.2
90° elbow	0.6	3	1.8
Transition (enlarging)	0.75	1	0.75
		Total ΣK	33.2

The dynamic head is $H_{\text{dyn}} = \dfrac{v_2{}^2 - v_1{}^2}{2g} + \left(\dfrac{fL}{d} + \Sigma K\right)\dfrac{v^2}{2g}$

Now $v_1 = 0$ (large tank), $v_2 = 0$ (included in spray head K factor),

$$f = 0.0223 \,, L = 70 \text{ m (total)}$$

Substituting:

$$H_{\text{dyn}} = 0 + \left(\frac{0.0223 \times 70}{0.125} + 33.2\right)\frac{v^2}{19.62}$$
$$= 2.328 \, v^2$$

Now $\qquad v = \dfrac{\dot{V}\,\text{m}^3/\text{s}}{A\ \text{m}^2} = \dfrac{\dot{V}}{A}\ \text{m/s}$

If the flow rate $\dot{V}$ is expressed in L/s, it is necessary to multiply by 10^{-3}.

That is, $\qquad v = \dfrac{\dot{V}}{A}\dfrac{\text{L/s} \times 10^{-3}}{\text{m}^2} = \dfrac{10^{-3}\,\dot{V}}{A}$

In this case the pipe has a diameter of 0.125 m

$$\therefore \ v = \frac{10^{-3}\,\dot{V}}{A(0.125)} = 0.0815\,\dot{V}$$
$$\therefore \ H_{\text{dyn}} = 2.328 \times (0.0815\,\dot{V})^2$$
$$= 0.0155\,\dot{V}^2$$

The static head is $H_{\text{stat}} = \dfrac{p_2 - p_1}{\rho g} + h_2 - h_1$

Now $\qquad p_2 = 20 \text{ kPa}, \ p_1 = 0, \text{ and } h_2 - h_1 = 6 \text{ m (total lift)}$

$$\therefore \ H_{\text{stat}} = \frac{20 \times 10^3 - 0}{992 \times 9.81} + 6$$
$$= 8.055 \text{ m}$$

Therefore, the system head equation (in terms of flow rate) is:

$$\boldsymbol{H = 8.055 + 0.0155\,\dot{V}^2}$$

(b) The system head curve may now be drawn by calculating the head at various flow
   rates (in the range 0–30 L/s). The values thus obtained are given in the table:

$\dot{V}$ (L/s)	0	5	10	15	20	25	30
H (m)	8.06	8.44	9.6	11.5	14.2	17.7	22

These points can now be plotted as shown in Figure 6.2:

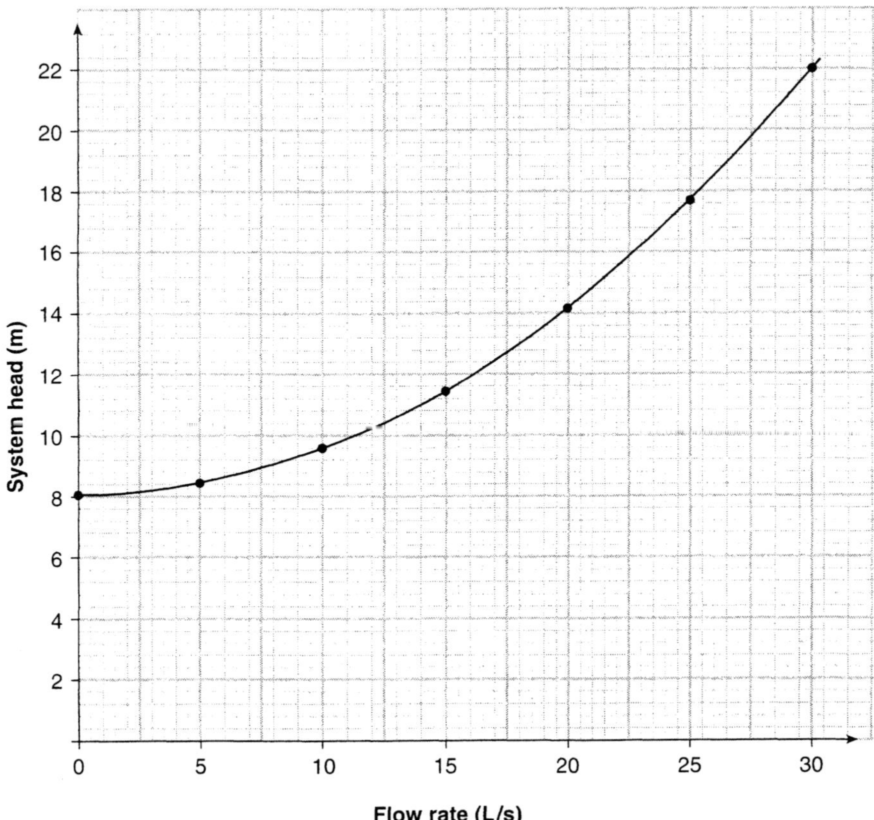

Fig. 6.2

6.2 *DUTY POINT*

When a pump is operating in a system with a steady flow rate of fluid there is equilibrium between the pump head–flow rate performance and the system head–flow rate characteristic. The equilibrium point is often termed the **duty point**. The location of the duty point for a system with a rotodynamic pump is shown graphically in Figure 6.3 on page 158.

For such a system the duty point can be located by superimposing the system head curve on the pump performance curve. Once the duty point has been found in this way the flow rate through the system and the pumping head can be read off. Also the efficiency, power and *NPSHR* can be read from the pump performance curves.

Example 6.2

The 125 × 80 pump whose performance curves are given in Appendix 12 is used in the system given in Example 6.1. Determine the duty point and the flow rate, head, power, efficiency and *NPSHR* for each impeller size.

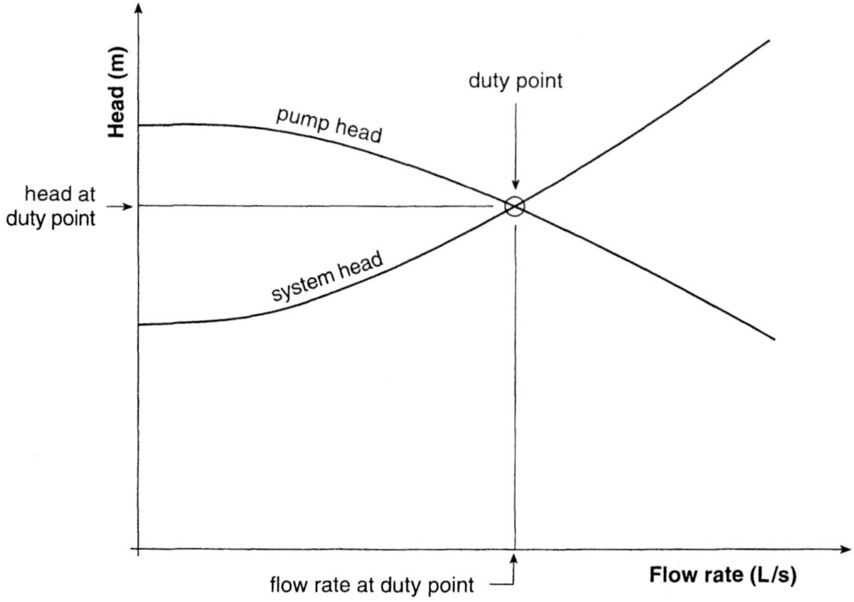

Fig. 6.3   *Duty point*

Solution

The system head curve is superimposed on the pump performance curves as shown in Figure 6.4 (opposite). That is, Figure 6.2 is redrawn on the same graph as the pump performance curves.

The duty point for each impeller size is marked on the graph, and for each duty point the following values may be read off:

Impeller size (mm)	Flow rate (L/s)	Head (m)	Power (kW)	Efficiency	NPSHR (m)
204	16.9	12.7	2.8	0.753	0.75
230	21.8	15.4	4.3	0.765	1.3
260	26.9	19.4	6.7	0.77	2.8

Note   Efficiency values are difficult to read exactly and require some creative interpolation.

Self-test problem 6.1

The pump whose performance curves are given in Appendix 11 is used to pump water in an 80 mm diameter system. The system head equation (in terms of velocity) is:

$$H = 4.5 + 0.39v^2$$

(a) Determine the duty point and the corresponding flow rate, head, power, efficiency and *NPSHR*.
(b) Check the efficiency value read graphically by calculation of the fluid power at the duty point.

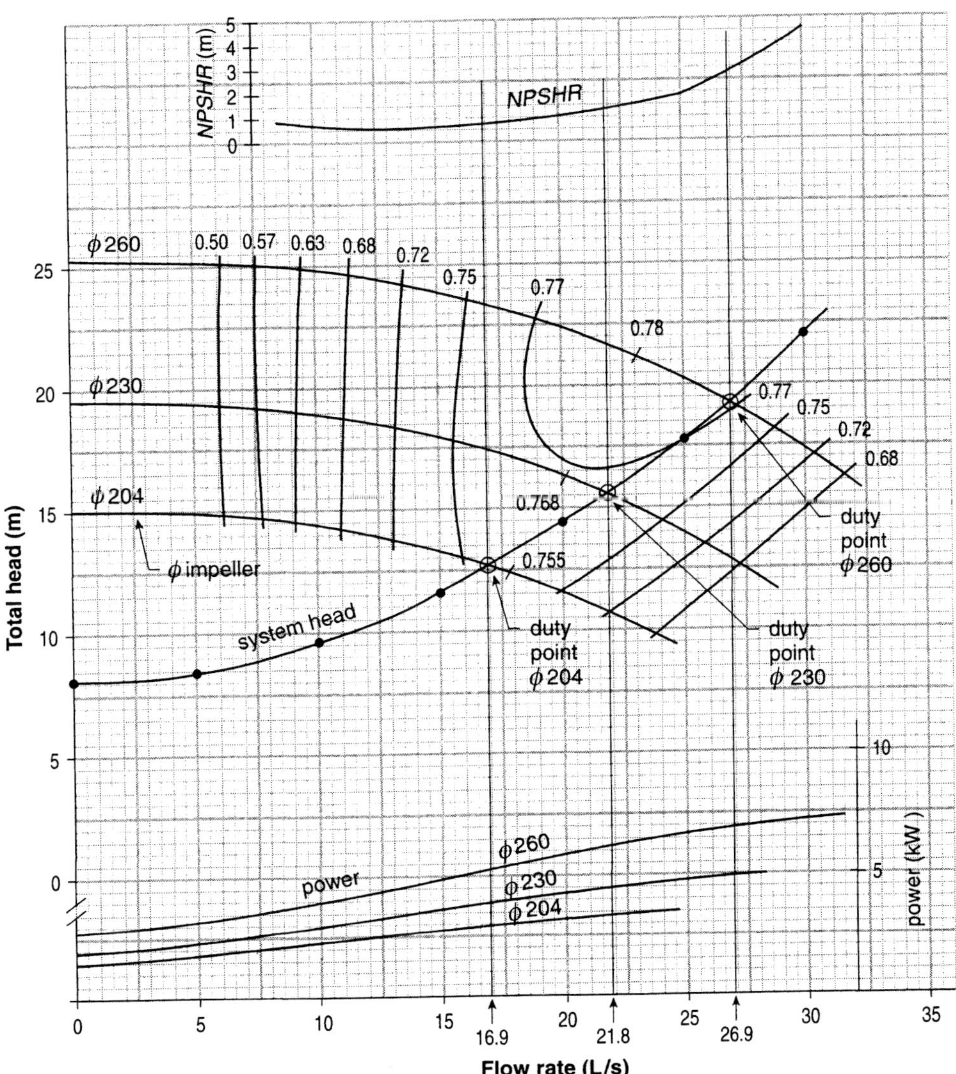

Fig. 6.4

6.3   *SELECTING THE TYPE OF PUMP TO BE USED IN A SYSTEM*

As described in Chapter 5, there are many types of pump available commercially, each with different characteristics. Selection of the best type of pump for a system may be dictated by overriding conditions such as priming capability, cleaning and hygiene requirements, the nature of the fluid, etc. If there are no requirements clearly dictating that one type is the best choice, calculation of the specific speed provides a guide to the most suitable type of pump for the required head, flow rate and shaft speed. That is, a pump operating within the appropriate specific speed range will give better efficiency than a pump operating outside this range. Put another way: if there are several types of pump that can provide the required flow rate and head, the best choice is the type that minimises the energy input (lowest energy cost).

Consider the system given in Example 6.1. Suppose that the required flow rate is about 20 L/s. Then at 20 L/s the system head (pumping head) is 14.2 m. Suppose also that a four-pole, directly coupled electric motor will be used to drive the pump so that the pump speed is about 1450 rpm.

$$\text{Now}\quad N_s = \frac{N\sqrt{\dot{V}}}{H^{\frac{3}{4}}} = \frac{1450 \times \sqrt{20}}{14.2^{\frac{3}{4}}} = 886$$

Clearly, this is in the range of specific speeds for which a centrifugal pump is most suitable. (See Figure 5.23 in Chapter 5.)

6.4 *SELECTING A ROTODYNAMIC PUMP*

If a rotodynamic pump is the best type of pump for the system under consideration, it is necessary to make a selection from those commercially available. There are two main reasons why this is so:

- The design and manufacture of a pump requires special expertise (technology) that has been acquired by specialist pump manufacturers over a long period of design and production.
- It is usually uneconomic to design and build a pump specially for a system unless there are large quantities involved. For example, motor engine manufacturers often design and manufacture the water pump used in the engine. In this case there are large numbers involved and the pump needs to be specially designed to fit into the block and to be driven by a belt drive and pulley system integral with the motor. However, this is a special case, and in most engineering applications it is more cost effective to purchase a stock pump.

To select a stock pump, it is necessary to refer to pump manufacturers' catalogues. These outline the dimensions and materials used in the range of pumps they produce. Also the performance curves for each type and size of pump are usually given. Rotodynamic pumps are often denoted by their inlet and outlet diameters and nominal impeller diameter. Using an example of this system of designation, a $100 \times 80 - 160$ pump would have a 100 mm inlet diameter, an 80 mm outlet diameter, and a nominal impeller diameter of 160 mm.

Initial selection of the appropriate pump is often assisted by use of a **range chart** which shows the operating range of the various sizes and types of pumps available. A typical range chart for a range of centrifugal pumps is shown in Figure 6.5 (opposite) (and also in Appendix 14).

Note   The upper set of curves is for two-pole speed (approx. 2950 rpm), whereas the lower set is for four-pole speed (approx. 1470 rpm) and also some at six-pole speed (approx. 960 rpm).

When using a range chart it is necessary to decide on the approximate flow rate and head needed in the system. For example, suppose that in the system given in Example 6.1 the required flow rate is about 20 L/s. From the system head equation, the system head at this flow rate is about 14 m. Using the range chart (Appendix 14) it can be seen that a $125 \times 80 - 250$ centrifugal pump running at four-pole speed is suitable for this application. This pump is manufactured with three standard impeller diameters: 260, 230 and 204 mm. It is now necessary to select which of these impellers is best. Usually there

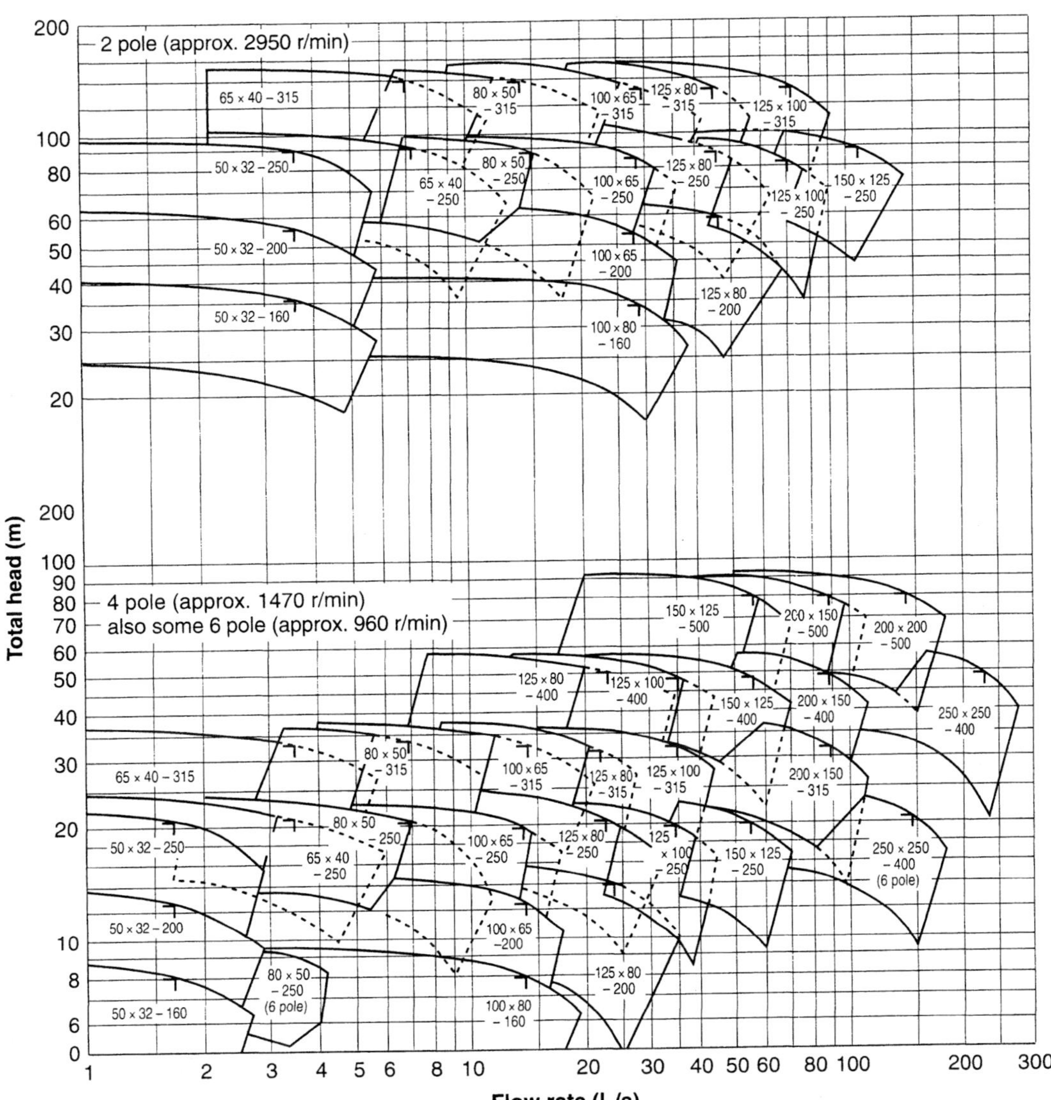

Fig. 6.5 *Range chart for centrifugal pumps*

is little (or no) difference in cost or size of the pump fitted with each different impeller size because all components (other than the impeller) are the same. The choice of impeller size is dictated primarily by energy requirements and the impeller that results in the lowest energy cost is usually the best choice.

It would appear logical to conclude that the impeller that has the highest operating efficiency will also have the lowest energy cost, but *this is not necessarily the case*. The reason is that the higher the flow rate at the duty point, the higher the flow loss. As a general rule it can be stated that *lowest energy cost occurs with the smallest pump or impeller that will give the required minimum performance*.[1] This rule is demonstrated in Example 6.3.

---

[1] This is similar to the case of a motor car where several engines of different size can be specified. Lowest consumption of fuel invariably occurs with the smallest engine size, regardless of engine efficiency.

Example 6.3

For the system given in Example 6.2 it is required that 800 m^3 of water be pumped per day. The pump will be driven by an electric motor which is 80% efficient. The cost of electrical energy is 10c per kWh. Determine the daily electrical energy cost of each impeller size and hence determine the impeller that is the best choice for lowest energy cost.

Solution

From Example 6.2 the flow rate and power of each pump is as tabulated below. Also the running time per day and the energy cost of each impeller size is calculated and shown in the table.

Impeller size (mm)	Flow rate (L/s)	Running time (h/day)	Shaft power (kW)	Electrical power (kW)	Electrical energy used per day (kWh)	Electrical energy cost per day
204	16.9	13.15	2.8	3.5	46.0	$4.60
230	21.8	10.2	4.3	5.375	54.8	$5.48
260	26.9	8.26	6.7	8.375	69.2	$6.92

Hence the rule applies and the smallest impeller, the 204 mm diameter one, uses the least amount of electrical energy and therefore has the lowest energy cost. This is despite the fact that this impeller has the *lowest efficiency* of the three sizes at the duty point. Therefore on the basis of least energy cost the **204 mm diameter impeller** is the best choice.

6.5 FLOW CONTROL USING A VALVE

Usually when a rotodynamic pump is used to pump fluid through a system, control of the flow rate is achieved by means of a valve (typically a globe valve). When the valve is wide open there is maximum flow through the system. If the valve is partly closed the head loss across the valve increases, that is, the K factor for the valve increases and the flow decreases. Consider the system head equation written in the form:

$$H = A + Bv^2$$

If there is no significant velocity head change between the inlet and outlet to the system (which is usually the case) factor B is given by:

$$B = \left(f\frac{L}{d} + \Sigma K \right) \times \frac{1}{2g}$$

An increase in the K factor of any fitting in the system increases the value of B. As B increases, the system head curve rises more steeply, so the duty point shifts to the left and occurs at a lower velocity and flow rate. This is shown in Figure 6.6 (opposite).

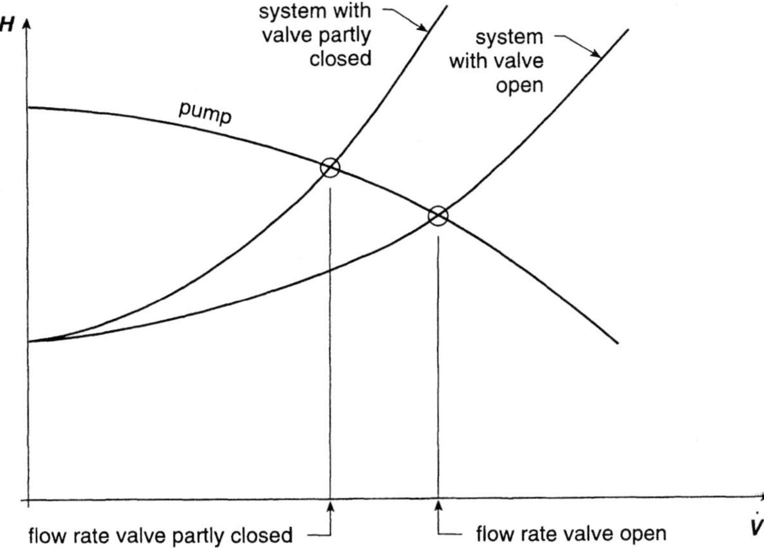

Fig. 6.6   *Effect of closing a valve in a system*

The use of a valve (manually or automatically operated) is a common and effective way of achieving flow control. The engineer can then specify a larger pump (or impeller size) than is actually needed in order to obtain the required flow rate. When the system is operational, the flow can be reduced to the required value by closing (throttling) the valve. This method provides maximum versatility and gives a safeguard against reduction in flow in the future due to filter clogging, sedimentation, corrosion, etc. However, there is also an increase in energy costs because energy is dissipated in friction and turbulence when a valve is closed.

The following generalisation can be made: For a certain flow rate through a system it is more efficient (less costly in energy) to use a smaller pump or impeller than a larger one where the flow rate is reduced by throttling a valve in the system. This is demonstrated in Self-test problem 6.2:

Self-test problem 6.2

The pump whose performance curves are given in Appendix 11 is used to pump water from a pond to a vented tank, the total increase in elevation being 3 m. The following data apply to the system:

- Pipe diameter 80 mm, length 20 m
- Friction factor 0.015
- ΣK = 9.5 (includes a fully open globe valve).

(a) Determine the flow rate, pump power and efficiency.
(b) If the globe valve is throttled down to the one-half open position, determine the new flow rate, pump power and efficiency.

6.6   *PLACEMENT OF A PUMP IN A SYSTEM*

Consider a system where a pump is used to pump liquid from one tank to another as shown in Figure 6.7 on page 164. Three possible locations for the pump are shown.

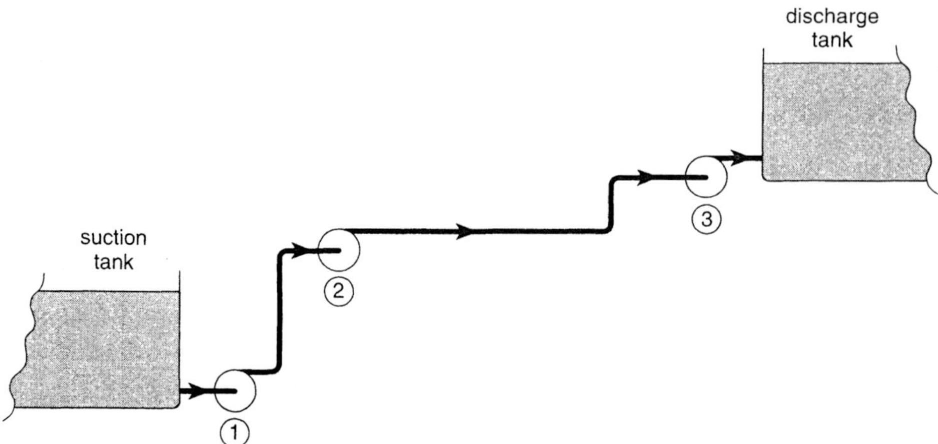

Fig. 6.7   *Three possible locations of a pump in a system*

In position ①, the pump is located as close as possible to the suction tank, so there is a small length of suction line and a large length of discharge line. The pump is below liquid level in the suction tank, so there is a positive static head at the pump inlet. That is, at low flow rates, the pressure at the pump inlet is above zero (gauge). At high flow rates, there will be a small head loss in the suction line, so there is a small reduction in head and pressure at the pump inlet, but these will remain positive.

In position ②, there is a longer length of suction line and a somewhat shorter length of discharge line. However, the pump is above liquid level in the suction tank, so there is a negative static head at the pump inlet. That is, at low flow rates, the pressure at the pump inlet is below zero (gauge). At high flow rates there will be a larger head loss in the suction line, and this will cause a further reduction in head and pressure at the pump inlet.

In position ③ the pump is located close to the delivery tank, so there is a long length of suction line and a short length of discharge line. The pump is well above liquid level in the suction tank, so there is a large negative static head at the pump inlet, that is, the pressure is well below zero (gauge). At high flow rates, there will be a large head loss in the suction line and this will cause a further large reduction in head and pressure at the pump inlet.

If the same diameter pipe is used on the inlet and outlet sides of the pump the head loss will be the same and the system head will be the same, regardless of the location of the pump. From this point of view all three positions are equivalent. However, several other important factors need to be considered: the necessity to prime and the possibility of cavitation.

Necessity to prime

If a rotodynamic pump is used, the pump is not self-priming, that is, it will not draw liquid and start pumping until the impeller is surrounded by liquid. The process of filling the pump casing with liquid on start-up is known as **priming** the pump.

In position ① priming is not necessary because there is already liquid in the pump. This is known as a flooded suction.

In positions ② and ③ priming is necessary whenever the pump is started up unless measures are taken to eliminate the need for priming after an initial priming (for the first time). Two such measures are illustrated in Figure 6.8 (opposite).

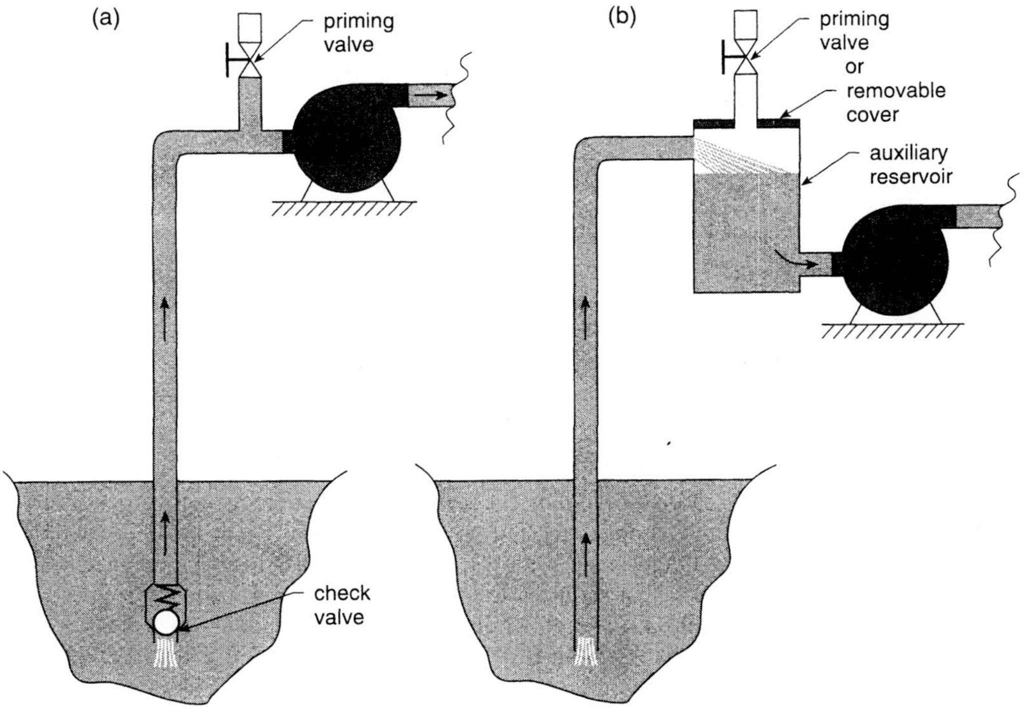

Fig. 6.8 *Two methods that avoid the need for priming on each start-up*

In method (a) a check valve or foot valve at the base of the suction line prevents liquid flowing back after the pump has been shut off. Initial priming is achieved by opening the priming valve and pouring liquid into the suction line. Thereafter the check valve prevents liquid flowing back and maintains liquid at the pump inlet. The disadvantage of this method is that the check valve can leak, particularly over an extended period of time, if a particle of foreign matter gets lodged between the valve and the seat.

In method (b) an auxiliary reservoir is installed at the pump suction. Initial priming is achieved by opening the priming valve (or cover to the auxiliary reservoir) and pouring liquid into it. Thereafter the reservoir remains full of liquid after the pump is shut down because liquid will not drain to a level below the level of the inlet pipe to the auxiliary reservoir.

Possibility of cavitation

Cavitation in a pump was discussed in Chapter 5 and it was seen that there is a minimum head required at the inlet of a pump above the vapour pressure head of the liquid in order to avoid cavitation. This is known as the net positive suction head required (*NPSHR*). The actual total head at the inlet of the pump above the vapour pressure head of the liquid is known as the net positive suction head available (*NPSHA*). The condition to avoid cavitation is:

$$NPSHA > NPSHR$$

Clearly, the greater the available head at the inlet to the pump, the less is the likelihood of cavitation. In order to maximise the available head at the pump inlet the head loss in the suction line should be kept as low as possible by ensuring the following:

- large suction pipe diameter;
- short length of suction pipe;
- smooth suction pipe;
- minimum of valves and fittings in the suction line (flow control valves should *never* be installed in the suction line, but always in the delivery line);
- fittings in the suction line chosen to minimise head loss (for example, long-sweep elbows);
- regular cleaning of filters and strainers in the suction line.

In addition, the liquid temperature should be as low as possible because the likelihood of cavitation increases with the saturation pressure of the liquid, and this increases with temperature. Finally the suction lift should be as small as practicable because the greater the suction lift, the lower is the pressure at the inlet to the pump.

Final choice of pump location

It is clear that on both grounds, avoiding the need to prime and reducing the possibility of cavitation, position ① is the best one. However, it is not always possible or feasible to install the pump below liquid level in the suction tank. A disadvantage of installing the pump below liquid level is that should a leak occur anywhere in the suction side, liquid will drain out of the reservoir. (The possibility of liquid draining out of the delivery tank can be readily overcome by ensuring that the delivery pipe is above liquid level in the delivery tank, or by installing check valves on the discharge side of the pump.)

Consequently position ② is often chosen for the location of the pump. Position ③ is highly undesirable (from a fluid mechanics point of view) and should be used only when other restraints make this the only possible location.

6.7 *CALCULATING THE NPSHA*

Whenever the pump inlet is above liquid level or there is a large head loss in the suction line it is necessary to ensure that:

$$NPSHA > NPSHR$$

The *NPSHR* is obtained from the pump performance curve, but the *NPSHA* needs to be determined by consideration of the suction side of the system. The calculation method is outlined below with reference to Figure 6.9 (opposite).

Applying the Bernoulli equation between ① and ② (and using absolute pressures for p_1 and p_2):

$$\frac{p_1}{\rho g} + \frac{v_1^2}{2g} + h_1 = \frac{p_2}{\rho g} + \frac{v_2^2}{2g} + h_2 + H_L$$

But $v_1 = 0$ (reservoir or tank), $h_1 = 0$ (datum), $h_2 = h$ and $v_2 = v$ (suction line velocity).

Substituting:

$$\frac{p_1}{\rho g} = \frac{p_2}{\rho g} + \frac{v^2}{2g} + h + H_L$$

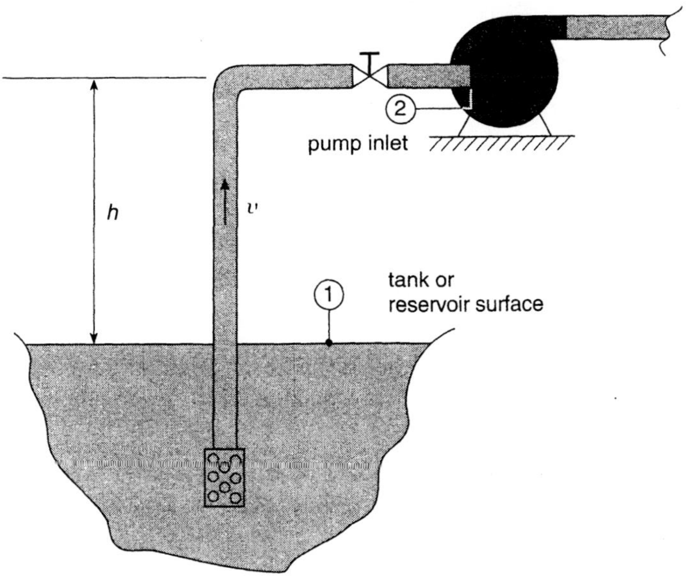

Fig. 6.9 *Typical pump inlet system*

But $p_2 = p_i$ (the pressure at the inlet to the pump)

$$\therefore \frac{p_i}{\rho g} = \frac{p_1}{\rho g} - \frac{v^2}{2g} - h - H_L \qquad \qquad \text{............................ (1)}$$

From the derivation of Formula 5.2 (given in Chapter 5):

$$NPSHA = \frac{p_i}{\rho g} + \frac{v^2}{2g} - \frac{p_v}{\rho g}$$

where p_i is the absolute pressure at the inlet to the pump, and p_v is the absolute saturation vapour pressure of the liquid being pumped.

Substituting $\dfrac{p_i}{\rho g}$ from Equation (1) into Formula 5.2, a formula for the net positive suction head available can be derived:

$$NPSHA = \frac{p_1}{\rho g} - \frac{v^2}{2g} - h - H_L + \frac{v^2}{2g} - \frac{p_v}{\rho g}$$

The $-\dfrac{v^2}{2g}$ and $+\dfrac{v^2}{2g}$ terms cancel so:

$$\boxed{NPSHA = \frac{p_1 - p_v}{\rho g} - h - H_L}$$ Net positive suction head available (6.1)

Formula 6.1 allows the *NPSHA* to be calculated for the system flow rate (at the duty point). Then knowing the *NPSHR* of the pump, the likelihood of cavitation can be ascertained.

Notes

- In Formula 6.1, p_1 is the absolute pressure at point ①. In Figure 6.1, point ① is at atmospheric pressure, so p_1 can be taken to have a value of 101.3 kPa (at sea level). If the suction tank is not vented and at a pressure other than atmospheric, the absolute value of this pressure should be used for p_1.
- Do not confuse p_1 and p_i as they are totally different. p_1 is the pressure (absolute) at the tank surface whereas p_i is the pressure (absolute) at the pump inlet.
- Formulas 5.2 and 6.1 are two formulas for calculating the *NPSHA* and both will produce the same answer. This is demonstrated in Self-test problem 6.3 below. However, Formula 6.1 is the most convenient to use because it does not require calculation of the suction line velocity.
- Appendix 13 may be used for the saturation vapour pressure (and density) of water at elevated temperatures. For other liquids use appropriate property tables.
- If the pump inlet is located below suction liquid level (flooded suction) Formula 6.1 becomes:

$$NPSHA = \frac{p_1 - p_v}{\rho g} + h - H_L$$

Example 6.4

Water at 30°C is pumped from a sump at atmospheric pressure, located 3.5 m below the pump inlet. At the duty point the head loss in the suction line is 1.4 m. Determine if cavitation is likely if the *NPSHR* at the duty point is 2.5 m.

Solution

From Appendix 13, for water at 30°C, $p_v = 4.25$ kPa and $\rho = 996$ kg/m³.
Substituting in Formula 6.1:

$$
\begin{aligned}
NPSHA &= \frac{p_1 - p_v}{\rho g} - h - H_L \\
&= \frac{101.3 \times 10^3 - 4.25 \times 10^3}{0.996 \times 10^3 \times 9.81} - 3.5 - 1.4 \\
&= 5.03 \text{ m}
\end{aligned}
$$

Since the *NPSHR* is 2.5 m, *NPSHA* > *NPSHR* and **cavitation is unlikely**.

# Self-test problem 6.3

A centrifugal pump draws water at 40°C from a sump at atmospheric pressure through an intake line 150 mm in diameter. At the duty point the flow rate is 35 L/s and the *NPSHR* is 2.2 m. The friction factor is 0.025 and the sum of the K factors for the fittings in the suction line is 15.5. The pump inlet is 2 m above the sump and the suction line is 1 m longer that the pump elevation. Use Appendix 13 for the saturation vapour pressure and density of water and:

(a) determine the pressure at the inlet to the pump in kPa (abs.);
(b) determine the *NPSHA* using Formula 5.2;
(c) determine the *NPSHA* using Formula 6.1 (it should give the same answer);
(d) determine the maximum elevation of the pump above the sump by assuming the same flow rate and head loss in the suction line;
(e) repeat (d) but take into account the additional head loss due to the increased length of suction pipe. Assume the same flow rate and the same number of fittings.

6.8 *SYSTEM WITH DIFFERENT DIAMETER SUCTION AND DISCHARGE LINES*

So far in this book, pumps have been used in systems with constant pipe diameter. However, many centrifugal pumps have an inlet diameter that is larger than the outlet diameter, and these pumps are installed in systems where the diameters of the suction and discharge pipes match the pump inlet and outlet diameters.

The main reason for using a larger diameter suction line than discharge line is that the larger diameter suction line produces less flow loss in the suction side of the system. In turn this means that the likelihood of cavitation is reduced. Because the suction line is usually of relatively short length and contains relatively few fittings, there is only a small cost and size penalty associated with a larger diameter suction line. Because the discharge line is often long and contains many fittings, it is often not cost effective or space effective to use a large-diameter pipe throughout. When there is a difference between the suction and discharge line diameters, a method of obtaining the system head and the duty point is as follows: Treat the suction and discharge sides initially as separate systems and obtain a system head equation for each side. Write these equations in terms of flow rate. The two equations can then be added to obtain a single equation for the system head. However, the *NPSHR* calculation is based on the suction side only.

The method is illustrated in Example 6.5.

Example 6.5

The pump whose performance curves are given in Appendix 11 is used to pump water at 25°C from an open tank to a closed tank under a pressure of 20 kPa (gauge) as shown in Figure 6.10.

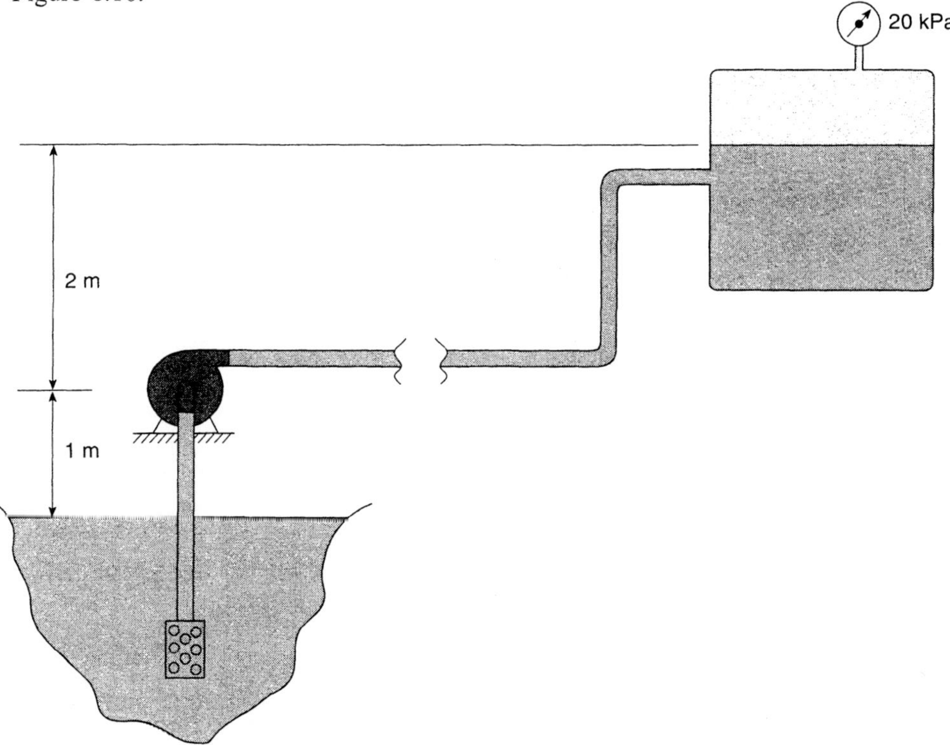

Fig. 6.10 *Figure for Example 6.5 (fittings not shown)*

The following data apply:

	Suction side	Discharge side
Pipe: drawn copper, diameter (mm)	100	80
length (m)	2.4	35
Fittings: gate valves (open)	1	2
globe valves (open)	—	1
90° elbows	1	3
hinged foot valve with strainer	1	—
sudden exit	—	1

The following assumption may be made: The friction factor has a constant value based on a flow rate of 12 L/s.

Write a system head equation in terms of flow rate in litres per second and determine the flow rate, head, power and efficiency at the duty point. Also check for cavitation by comparing the *NPSHA* with the *NPSHR*.

Solution

From Appendix 13, for water at 25°C: $\rho = 997$ kg/m^3 and $\mu = 0.9 \times 10^{-3}$ Pas
From Appendix 4: $\varepsilon = 1.5 \times 10^{-3}$ mm
From Appendix 6: the K factor table is as follows:

Fitting	Suction side	Discharge side
Gate valve (open)	1 × 0.2 = 0.2	2 × 0.2 = 0.4
Globe valve (open)	—	1 × 6.0 = 6.0
90° elbow	1 × 0.6 = 0.6	3 × 0.6 = 1.8
Hinged foot valve with strainer	1 × 3.0 = 3.0	—
Sudden exit	—	1 × 1.0 = 1.0
Total ΣK	3.8	9.2

For a flow rate of 12 L/s, the friction factor can be calculated as follows:

	Suction side	Discharge side
$v = \dfrac{\dot{V}}{A}$ (m/s)	1.528	2.387
$Re = \dfrac{v d \rho}{\mu}$	0.17×10^6	0.212×10^6
$\varepsilon_R = \dfrac{\varepsilon}{d}$	0.015×10^{-3}	0.0188×10^{-3}
$f = 0.0055\left[1 + \left(20000\varepsilon_R + \dfrac{10^6}{Re}\right)^{\frac{1}{3}}\right]$	0.0156	0.0150

There is no velocity head, so the dynamic head is the head loss:

$$H_L = \left(f\frac{L}{d} + \Sigma K \right) \frac{v^2}{2g}$$

Suction side:

$$H_L = \left(0.0156 \times \frac{2.4}{0.1} + 3.8 \right) \frac{v^2}{19.62} = 0.2128 v^2$$

When $\dot{V}$ is in L/s, $v = \dfrac{10^{-3}}{A(0.1)} \dot{V} = 0.1273\,\dot{V}$

$$\therefore H_L = 0.2128 \times (0.1273\,\dot{V})^2 = 0.00345\,\dot{V}^2$$

Discharge side:

$$H_L = \left(0.015 \times \frac{35}{0.08} + 9.2 \right) \frac{v^2}{19.62} = 0.8025 v^2$$

When $\dot{V}$ is in L/s, $v = \dfrac{10^{-3}}{A(0.08)} \dot{V} = 0.199\,\dot{V}$

$$\therefore H_L = 0.8025 \times (0.199\,\dot{V})^2 = 0.0318\,\dot{V}^2$$

Therefore the total head loss for the suction and discharge sides is:

$$H_L = 0.00345\,\dot{V}^2 + 0.0318\,\dot{V}^2 = 0.0352\,\dot{V}^2$$

The total static head is:

$$h + \frac{p}{\rho g} = 3 + \frac{20 \times 10^3}{997 \times 9.81} = 5.04 \text{ m}$$

Therefore the system head equation is $H = 5.04 + 0.0352\,\dot{V}^2$

The system head can now be calculated for various flow rates as shown in the table:

$\dot{V}$ (L/s)	0	2	4	6	8	10	12	14
H (m)	5.04	5.2	5.6	6.3	7.3	8.6	10.1	11.9

These points can now be plotted on the pump performance curve as shown in Figure 6.11 on page 172.

At the duty point, $\dot{V} = \mathbf{10.4}$ **L/s,** $H = \mathbf{8.9}$ **m,** $P = \mathbf{1.22}$ **kW,** $\eta = \mathbf{73.5\%}$
To calculate the *NPSHA* it is necessary to calculate the head loss in the suction line. As determined above, this is:

$$H_L = 0.00345\,\dot{V}^2$$

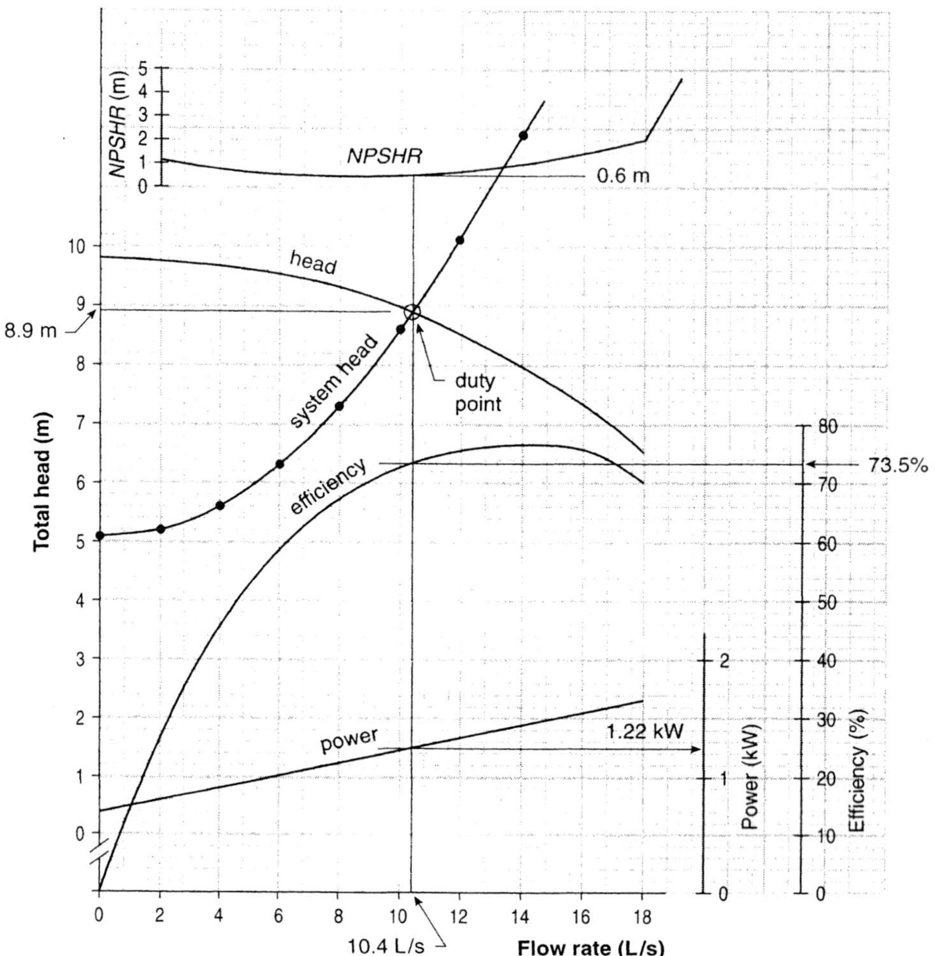

Fig. 6.11

At the duty point the flow rate is 10.4 L/s and the head loss in the suction line is:

$$H_L = 0.00345 \times 10.4^2 = 0.373 \text{ m}$$

From Appendix 13, for water at 25°C, p_v = 3.18 kPa (abs.)
Using Formula 6.1:

$$NPSHA = \frac{p_1 - p_v}{\rho g} - h - H_L$$

On the suction side, h = 1 m and p_1 = 101.3 kPa (abs.)

Substituting:

$$NPSHA = \frac{101.3 \times 10^3 - 3.18 \times 10^3}{0.997 \times 10^3 \times 9.81} - 1.0 - 0.373$$

$$= 10.03 - 1.0 - 0.373$$

$$= \mathbf{8.66 \ m}$$

From Figure 6.11, at the duty point, $NPSHR = \mathbf{0.6 \ m}$
Therefore $NPSHA > NPSHR$ and **cavitation is unlikely**

6.9 *SELECTING A POSITIVE-DISPLACEMENT PUMP*

If the special requirements of the system, or the value of the specific speed, are such that a positive-displacement pump is the best choice, selection of the pump follows essentially the same procedure as for selection of a rotodynamic pump. However, the flow rate of a positive-displacement pump is less dependent on system head (or pressure) than is the case with a rotodynamic pump. Rather, the flow rate through a positive-displacement pump depends primarily on the speed of the pump.

If performance curves are available, they may be used in a similar way to those provided with rotodynamic pumps to determine the flow rate and other operating parameters when the pump is installed in a system. If performance curves are not available, theoretical performance can be calculated and operating efficiencies can be estimated. When these efficiencies are applied to the theoretical values a close estimate of the actual performance can be obtained.

The likelihood of cavitation can be checked in the same way as for a rotodynamic pump by calculating the *NPSHA* and comparing it to the *NPSHR*. The only difference is that positive-displacement pumps often produce a pulsating flow and in such cases the acceleration head (h_A) needs to be deducted when calculating the *NPSHA*. That is, Formula 6.1 becomes:

$$NPSHA = \frac{p_1 - p_v}{\rho g} - h - H_L - h_A$$

The acceleration head depends primarily on the impulsiveness of the pump, but a number of other factors such as the length of suction line, the suction line velocity and the speed of the pump have a bearing on it. The impulsiveness of a pump varies with the number of chambers. For example, the more cylinders there are in a reciprocating piston pump, the lower the impulsiveness. However, calculation of the acceleration head is beyond the scope of this book.

Example 6.6

The gear pump whose performance curves are given in Appendix 10 is driven at a speed of 1450 rpm and used in a hydraulic system where the delivery pressure is 2 MPa. Determine:
(a) the flow rate, power and efficiency;
(b) the fluid power, and hence check the efficiency value read off the chart.

If this pump is run at a speed of 950 rpm with the same delivery pressure and efficiency, determine:
(c) the new flow rate and power.

Solution

(a) Referring to Appendix 10, reading off at 2 MPa pressure as shown in Figure 6.12:

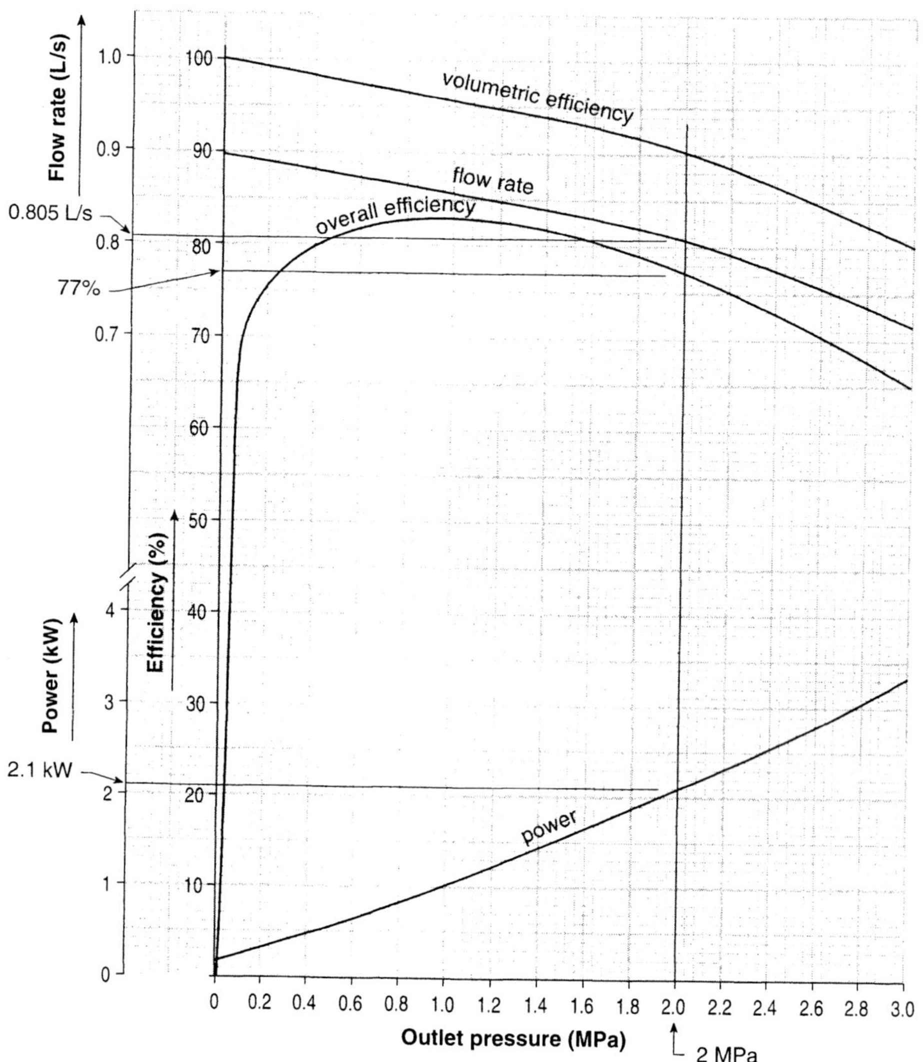

Fig. 6.12

$$\dot{V} = \textbf{0.805 L/s}, P = \textbf{2.1 kW}, \eta = \textbf{77\%}$$

(b) $P_f = p\dot{V} = 2 \times 10^6 \times 0.805 \times 10^{-3} = 1610\ \text{W} = 1.61\ \text{kW}$

$$\eta = \frac{P_f}{P} = \frac{1.61}{2.1} = \textbf{77\%}\ \text{(Checks)}$$

(c) If the volumetric efficiency is the same, the flow rate will be directly proportional to the speed.

$$\therefore \text{at 950 rpm, } \dot{V} = \frac{950}{1450} \times 0.805 = \textbf{0.527 L/s}$$

If the overall efficiency is the same and the delivery pressure is the same, then the power will be directly proportional to the speed.

$$\therefore \text{ at 950 rpm, } P = \frac{950}{1450} \times 2.1 = \mathbf{1.38 \text{ kW}}$$

Self-test problem 6.4

A six-cylinder, variable-volume piston pump is to be used in a hydraulic application where the pressure is 4 MPa. The maximum flow rate is to be 0.5 L/s. The pump is to be run at a speed of 960 rpm. The volumetric efficiency is 90% and the overall efficiency is 80%. At maximum flow rate the bore and stroke are the same. Determine:
(a) the required bore size;
(b) the input power at maximum flow rate.

Summary

When a fluid is pumped through a system, the system head increases with flow rate as a parabolic (or nearly parabolic) curve. There is a static component that does not vary with flow rate, and a dynamic component that varies as the square of the flow rate. The dynamic component is essentially the head loss but sometimes there is also a velocity head component. The pump used in the system has a performance characteristic that shows the relationship between the total head delivered by the pump and the flow rate. For example, a centrifugal pump usually has a falling characteristic, that is, the total head decreases as flow rate increases.

When a pump is installed in a system, equilibrium is achieved between the pump head–flow rate performance and the system head–flow rate demand. The point of equilibrium is often known as the duty point. It can be determined graphically by superimposing the system head–flow rate curve on the pump performance curve. The duty point is where the two curves intersect. Once the duty point has been located, the flow rate through the system can be determined, as well as the head, efficiency, power and *NPSHR*.

When selecting the best type of pump to be used in a system, selection may be dictated by special requirements. For example, if cleanliness and freedom from contamination are essential, a peristaltic pump may be the only suitable choice. If there are no special requirements of this type, the specific speed provides a useful guide to the most suitable type of pump for the desired flow rate and head at the available shaft speed. As a general rule, a pump operating within the appropriate specific speed range for the pump type (see Figure 5.23, Chapter 5) will give better efficiency than one operating outside this range.

If a rotodynamic pump is the best type for the system under consideration, it is necessary to select the most appropriate type from those commercially available. Initial selection is often assisted by reference to a range chart which shows the operating range for the various types and sizes of pump.

In the case of centrifugal pumps there are often a number of impeller diameters available for each pump. Final selection of the best impeller diameter is usually based on operating economy, that is, the best choice is the impeller that gives the required performance at the lowest energy cost. It is not necessarily the case that the impeller with the highest efficiency at the duty point has the lowest energy cost. Rather, lowest input of energy is generally achieved by the smallest impeller that gives the required minimum performance.

An option available to the engineer is to specify a larger pump or impeller than needed for the design flow rate, and then provide a flow control valve so that flow can be throttled down until the required flow rate is achieved. This option gives maximum versatility, and provides a safeguard against future flow reduction due to filter clogging or growth. or sedimentation, but it is also costly in energy because the flow loss across the throttled flow control valve is a non-recoverable energy loss.

The location of the pump in the system is dictated by two main considerations (other than convenience of installation). These are the possibility of cavitation, and the need to prime. As a general rule it is best to ensure that the suction side of the system is of minimum length and that the suction lift is a minimum. This reduces the risk of cavitation. If the pump is located below suction liquid level there is a flooded inlet, and a rotodynamic pump will not need to be primed. The likelihood of cavitation can be reduced by minimising flow losses in the suction line by use of a large-diameter, smooth pipe and a minimum number of fittings. Flow control valves should never be located in the suction line because cavitation will inevitably occur if a valve in the suction line is throttled down.

The possibility of cavitation can be checked by calculating the *NPSHA* and making sure that *NPSHA* > *NPSHR*. For a rotodynamic pump, *NPSHA* can be calculated using Formula 6.1:

$$NPSHA = \frac{p_1 - p_v}{\rho g} - h - H_L$$

where p_1 is the absolute pressure at the suction liquid surface, p_v is the absolute vapour pressure of the liquid (at the given temperature), h is the suction lift, and H_L is the head loss in the suction line.

It is often the case that different diameter suction and discharge pipes are used, with the suction line of greater diameter than the discharge line. When this is the case, each side may be treated as a separate system when calculating the friction factor and head loss. If the dynamic head is expressed in terms of flow rate on each side of the system (rather than velocity) the two terms can be added to obtain a single expression for the total dynamic head (as a function of flow rate). The total system static head can then be added to the total dynamic head and a single equation obtained for the total system head. This equation can be treated in the same way as for a single-diameter system in order to find the duty point. That is, the total system head curve can be drawn and superimposed on the graph of the pump performance curve. However, the suction side needs to be treated separately when calculating the *NPSHA*.

Selection of a positive-displacement pump follows essentially the same procedure as for rotodynamic pumps. If performance curves are not available, the performance can be estimated theoretically and modified by use of suitable efficiency factors. The *NPSHA* calculation is the same, except that the acceleration head needs to be deducted in Formula 6.1.

 # Problems

Notes

- Use the Appendixes for any required data not given.
- Assume a constant friction factor unless stated otherwise.

6.1 **(a)** With the aid of a neat graph, explain what is meant by *duty point*.

(b) What is the advantage of operating a pump within the appropriate specific speed range as shown in Figure 5.23 (Chapter 5)?

(c) State two reasons why it is generally best for an engineer to select a commercially available pump rather than to design a special pump.

6.2 (a) With the aid of a neat graph, explain the effect of closing a valve in a fluid pumping system.

(b) Where should a flow control valve be located in a fluid pumping system, and why?

(c) What are the advantages and disadvantages of selecting a larger pump than necessary for a fluid pumping system, and then throttling the flow back to the desired value using a flow control valve?

6.3 Draw a neat diagram of a system where liquid is pumped from one tank to another. Show three possible locations for the pump as follows :
(1) Close to the suction tank and below liquid level in the suction tank.
(2) Close to the suction tank and above liquid level in the suction tank.
(3) Close to the delivery tank.
Discuss the merit of each of these locations with regard to:
(a) the necessity to prime;
(b) the possibility of cavitation.

6.4 (a) What is meant by priming a pump, and how is this usually accomplished?

(b) What group of pumps usually does not require priming, and what group does?

(c) With the aid of a neat sketch, discuss briefly two ways of overcoming the need to prime after an initial prime (for the first time) when a pump is used that is not self-priming.

6.5 Briefly discuss ways of minimising the risk of cavitation, under the headings:
(a) Head loss in the suction line;
(b) Liquid temperature;
(c) Suction lift.

6.6 (a) Why is it undesirable to locate a flow control valve on the suction side of a pump?

(b) Name two types of valve that are often located on the suction side of a pump, and state their purpose.

(c) Explain what is meant by the impulsiveness of a pump, and name a pump that has a high impulsiveness.

(d) How does impulsiveness affect the possibility of cavitation?

6.7 Use the range chart (Appendix 14) to select a centrifugal pump for systems where the required flow rate and head are as given in the following table. State the electric motor pole type needed to drive the pump.

	Flow rate (L/s)	*Head* (m)
(a)	3	30
(b)	3	80
(c)	20	10
(d)	20	50
(e)	50	70

(a) 65 × 40 – 315, 4 pole (b) 50 × 32 – 250, 2 pole (c) 125 × 80 – 200, 4 pole (d) 100 × 65 – 200, 2 pole or 125 × 100 – 400, 4 pole (e) 125 × 80 – 250, 2 pole or 150 × 125 – 500, 4 pole

6.8 In a chilled-water system for an airconditioning plant, the flow rate of water is controlled to 20 L/s when the airconditioning system is operating. At this flow rate the system head with all valves open is 10 m. The 125 × 80 − 250 pump whose performance curves are given in Appendix 12 is used and the flow rate is automatically adjusted by means of a valve on the discharge side of the pump. The plant operates for an average of 10 hours per day for 150 days per annum. Electricity costs an average of 10.3c per kWh and the motor driving the pump has an efficiency of 87%.

Determine the shaft power and annual cost of electricity (to the nearest dollar) associated with each of the three impeller sizes available for this pump.

ϕ 204: 3.1 kW, $551 ϕ 230: 4.2 kW, $746 ϕ 260: 5.75 kW, $1021

6.9 A centrifugal pump used in a system has a head–flow rate performance as given in the following table:

Flow rate (L/s)	0	10	20	30	40	50
Head (m)	60	58	55	51	46	40

The pump is used in a system where the system head equation is:

$$H = 16.5 + 0.018\,\dot{V}^2 \ (H \text{ in m and } \dot{V} \text{ in L/s})$$

Determine the flow rate through the system.

40.2 L/s

6.10 The pump whose performance curves are given in Appendix 11 is used to pump water from a pond to a vented tank 3 m above it. The pipe is 80 mm in diameter and of total length 30 m. The friction factor has an average value of 0.017 and the sum of the K factors in the system is 12.5 when all valves are fully open.
(a) Write the system head equation in terms of velocity (in m/s).
(b) Write the system head equation in terms of flow rate (in L/s).
(c) Determine the flow rate, head, power and efficiency at the duty point.
(d) Calculate the fluid power at the duty point, and hence check the efficiency value read graphically.

(a) $H = 3 + 0.962v^2$ (b) $H = 3 + 0.0381\,\dot{V}^2$ (c) 12 L/s, 8.5 m, 1.34 kW, 75%
(d) 1.0 kW, 75%

6.11 Repeat Problem 6.10 if a globe valve in the system is now closed down to the one-quarter open position.

(a) $H = 3 + 1.88v^2$ (b) $H = 3 + 0.0744\,\dot{V}^2$ (c) 9.2 L/s, 9.1 m, 1.15 kW, 71%
(d) 0.821 kW, 71.4%

6.12 A pressure gauge at the inlet to a pump reads −20 kPa when pumping water at 30°C with velocity of 1.5 m/s in the suction line. Determine the *NPSHA*.

8 m

6.13 A centrifugal pump draws water at 30°C from a sump, the water level being 2 m below the pump inlet. The head loss in the suction line is 1.5 m.
(a) Determine the *NPSHA* if the sump is at atmospheric pressure.
(b) Determine the maximum allowable vacuum in the sump if the *NPSHR* is 2.4 m.

(a) 6.43 m (b) −39.4 kPa

6.14 A hot-water pump, pumping water at 60°C, is used in a plant located on a mountain. The water is drawn from a vented tank whose surface is 3 m below the pump inlet. The suction line velocity is 2 m/s and the head loss in the suction line is 1.2 m. A mercury barometer in the plant shows a height of 680 mm. The relative density of mercury can be taken as 13.6. Determine the net positive suction head available.

> 3.13 m

6.15 A centrifugal pump is used to pump hot water at 80°C from a solar pond to an absorption refrigeration plant. The suction line is 200 mm in diameter and the pump inlet is 1.3 m above the surface of the pond. The suction pipe is 2 m long, made of drawn copper, and contains a hinged foot valve with strainer, a 90° elbow, and an open gate valve. When the flow rate is 60 L/s, determine:
(a) the friction factor;
(b) the head loss in the suction line;
(c) the pressure at the pump inlet in kPa (abs.);
(d) the *NPSHA* using Formula 5.2;
(e) the *NPSHA* using Formula 6.1;
(f) the maximum elevation of the pump inlet above the pond if the *NPSHR* is 1.8 m. Assume the same flow rate and head loss in the suction line.

> (a) 0.0112 (b) 0.727 m (c) 80.2 kPa (d) 3.63 m (e) 3.63 m (f) 3.13 m

6.16 In a mobile fire-fighting truck as depicted in Figure P6.16, water at 20°C is pumped through a hose of diameter 80 mm and total length 100 m to the nozzle located 20 m above the water level in the vented tank. The absolute roughness of the hose is 0.01 mm. Both valves are fully open. The nozzle reduces the water stream to a diameter of 30 mm at the exit of the nozzle. The nozzle is gradually tapered, so the K factor is negligible, but the velocity head of the water at the exit from the nozzle is appreciable and *must* be taken into account. Use a constant friction factor based on a flow rate of 15 L/s. Determine:
(a) the system head equation in the form: $H = A + Bv^2$ (H in m and v in m/s);
(b) the system head equation in the form: $H = A + C\dot{V}^2$ (H in m and $\dot{V}$ in L/s)
(c) the system head at the flow rates set out in the table below:

Flow rate (L/s)	0	5	10	15	20	25
System head (m)						

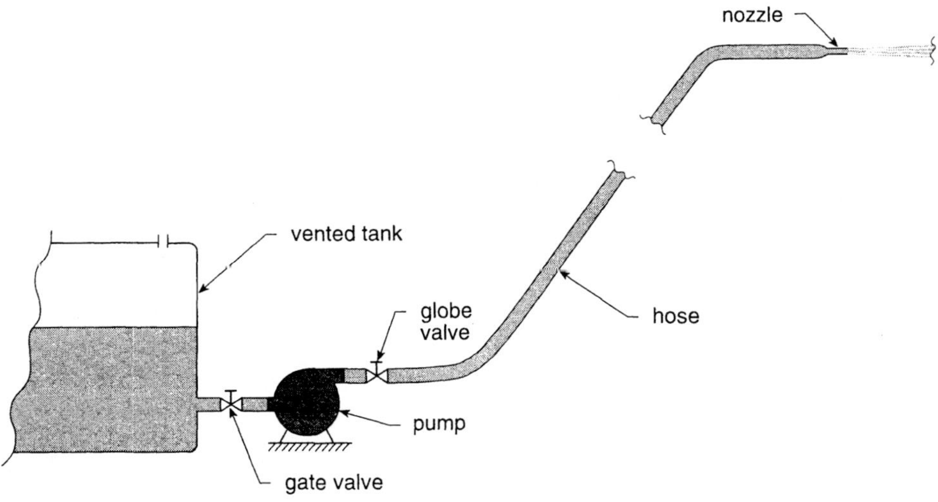

Fig. P6.16

(a) $H = 20 + 3.93v^2$ (b) $H = 20 + 0.1555\dot{V}^2$

(c)

Flow rate (L/s)	0	5	10	15	20	25
System head (m)	20	23.9	35.6	55.0	82.2	117

6.17 In the mobile fire-fighting truck system given in Problem 6.16, the $125 \times 80 - 260$ pump whose performance curves are given in Appendix 12 is to be used. However, these curves relate to the pump running at a nominal speed of 1470 rpm (four-pole electric motor speed), whereas the pump will actually run at 2500 rpm using power taken off the main engine. Redraw the pump head–flow rate curve for this speed, and hence determine the flow rate.

18.2 L/s

6.18 In a fluid pumping system, fluid with relative density 0.78 and saturation vapour pressure 12.5 kPa (abs.) is pumped from one tank to another. The following data apply:

	Suction side	Discharge side
Tank pressure (kPa (g))	0 (atm)	30
Static lift (m)	2.5	7.8
ΣK (including entrance or exit)	8.5	28.2
Pipe diameter (mm)	150	150
Pipe length (m)	4.5	34
Friction factor (assumed constant)	0.018	0.018

The system head equation is of the form:

$$H = A + C\dot{V}^2 \quad (H \text{ in m and } \dot{V} \text{ in L/s})$$

(a) Determine the value of the factor C for the suction side only.
(b) Determine the value of the factor C for the discharge side only.
(c) Write the system head equation for the whole system.
(d) Complete the following table:

Flow rate (L/s)	0	10	20	30	40	50
System head (m)						

(e) Determine the maximum flow rate to avoid cavitation if the *NPSHR* is 2.0 m.

(a) 1.476×10^{-3} (b) 5.269×10^{-3} (c) $H = 14.22 + 6.744 \times 10^{-3}\dot{V}^2$

(d)

Flow rate (L/s)	0	10	20	30	40	50
System head (m)	14.2	14.9	16.9	20.3	25.0	31.1

(e) 69.4 L/s

6.19 For the system given in Problem 6.18, suppose that it is decided to use 125 mm diameter pipe instead of 150 mm diameter on the discharge side. The friction factor on the discharge side will now be 0.015 and all other data will be the same. Repeat (b), (c) and (d).

(b) 0.0109 (c) $H = 14.22 + 0.0124\,\dot{V}^2$

(d)

Flow rate (L/s)	0	10	20	30	40	50
System head (m)	14.2	15.5	19.2	25.4	34.1	45.2

6.20 The gear pump whose performance curves are given in Appendix 10 is used in a hydraulic system where the oil has a relative density of 0.85 and a viscosity of 0.2 Pas. The oil is pumped from a storage tank at atmospheric pressure through a 30 mm diameter steel pipe of length 20 m to a hydraulic accumulator. The pump outlet pressure is 1.5 MPa(g). There is negligible elevation change between the storage tank and the accumulator. The total K factor for the system is 25.5. Determine:
(a) the flow rate of oil;
(b) the input power;
(c) the efficiency;
(d) the fluid power and efficiency using this value of the fluid power;
(e) the friction factor;
(f) the head loss in m;
(g) the pressure drop in kPa due to the head loss;
(h) the pressure at the accumulator.

(a) 0.835 L/s (b) 1.55 kW (c) 81% (d) 1.25 kW, 80.8% (e) 0.109 (f) 6.98 m
(g) 58.2 kPa (h) 1442 kPa

6.21 In an irrigation application as shown in Figure P6.21, bore water at 20°C is pumped from a bore located 4 m below the pump inlet through a pipe to a sprinkler.

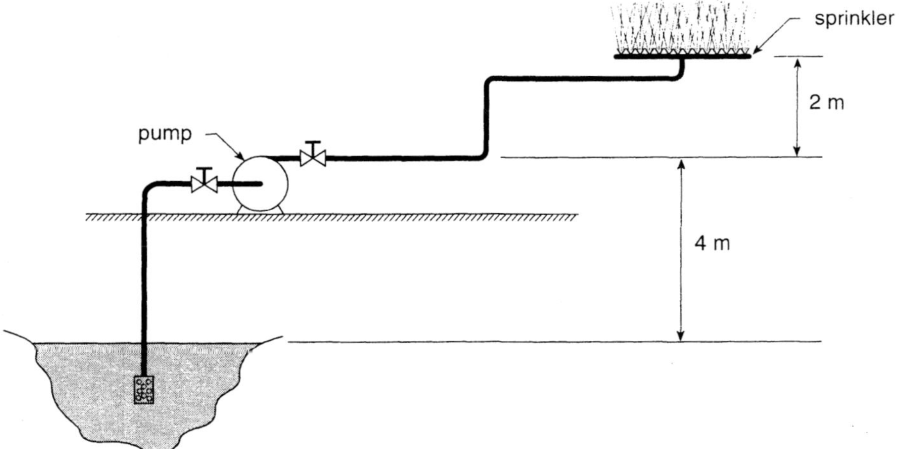

Fig. P6.21

Galvanised steel pipe of diameter 30 mm is used throughout; the length on the suction side is 5 m and on the discharge side 80 m. The rotating sprinkler has 100 holes and the water leaves them in jets of diameter 1.5 mm. The following fittings are used:

Fitting	Suction side	Discharge side
Hinged foot valve with strainer	1	—
Screwed socket	3	20
90° elbow	1	2
Open gate valve	1	1
Sprinkler	—	1

The K factor for the sprinkler is 0.5 (but this does *not* include an allowance for the exit loss, that is, the loss of velocity head after the water leaves the sprinkler). A positive-displacement pump is to be used and the design flow rate is 2 L/s. For this flow rate, determine:

(a) the gauge pressure at the pump inlet;
(b) the gauge pressure at the pump outlet;
(c) the pressure increase across the pump;
(d) the power required to drive the pump if the pump efficiency is 80%;
(e) the *NPSHA* if the acceleration head is 1.0 m;
(f) the daily energy cost of running the pump for 8 hours per day, if the electric motor driving the pump is 85% efficient and electricity costs 11c per kWh.
(a) −80.0 kPa (b) 430.3 kPa (c) 510.3 kPa (d) 1.276 kW (e) 1.345 m (f) $1.32

6.22 For the irrigation system given in Problem 6.21, determine:
(a) the system head (pumping head) at the design flow rate;
(b) the system head equation in the form: $H = A + Bv^2$ (H in m and v in m/s);
(c) the system head equation in the form: $H = A + C\dot{V}^2$ (H in m and $\dot{V}$ in L/s);
(d) the system head for flow rates as follows:

Flow rate (L/s)	0	1	1.5	2	2.5	3
System head (m)						

(e) the flow rate through the system if a centrifugal pump whose head–flow rate performance is as follows is used (rather than a positive-displacement pump);

Flow rate (L/s)	0	1	2	3	4	5
System head (m)	87	86	84	81.5	78	72.5

(f) the pump designation and nominal speed, using the range chart (Appendix 14).
(a) 52.12 m (b) $H = 6 + 5.762v^2$ (c) $H = 6 + 11.53\dot{V}^2$
(d)

Flow rate (L/s)	0	1	1.5	2	2.5	3
System head (m)	6.0	17.5	32.0	52.1	78.1	110

(e) 2.6 L/s (f) 50 × 32 − 250, 2950 rpm (2 pole)

6.23 Water at 15°C is pumped through the system shown in Figure P6.23:

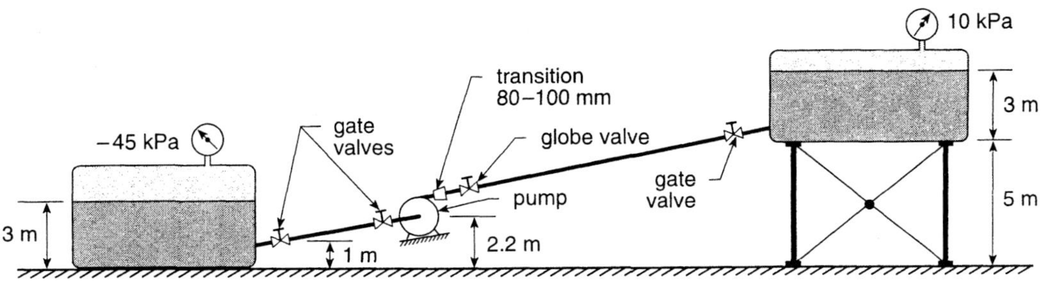

Fig. P6.23

The performance curves for the pump are as given in Appendix 12. The pipe is drawn copper of diameter 125 mm and length 24 m on the suction side, and diameter 100 mm and length 56 m on the discharge side. All valves are open. Use a friction factor based on a flow rate of 20 L/s. Determine:

(a) the system head equation in terms of flow rate in L/s.

Also for each impeller size, determine:

(b) the flow rate through the system;

(c) the *NPSHA* and *NPSHR*;

(d) the efficiency;

(e) the input power;

(f) the electricity cost per day in order to pump 500 m³ of water from the lower tank to the upper tank. Electricity is charged at 9.5c per kWh and the electric motor efficiency is 85%.

If the globe valve is throttled down the half-open position, determine:

(g) the new system head equation in terms of flow rate in L/s;

(h) the new flow rate with each impeller size.

(a) $H = 10.61 + 0.01472\,\dot{V}^2$ (b)–(f) see table (g) $H = 10.61 + 0.0197\,\dot{V}^2$ (h) see table

	φ 204	φ 230	φ 260
(b) $\dot{V}$ (L/s)	14.3	19.7	25.3
(c) *NPSHA* (m)	6.1	5.86	5.53
NPSHR (m)	0.5	1.0	2.0
(d) η (%)	73.5	76.6	77.4
(e) *P* (kW)	2.6	4.1	6.6
(f) Cost in $/day	2.82	3.23	4.05
(h) $\dot{V}$ (L/s)	12.7	18.0	23.0

Solutions to self-test problems

Chapter 1

1.1 **(a)** $\dot{m} = vA\rho$. $\therefore 1.2 = v_1 \times A(0.025) \times 800$. $\therefore v_1 = $ **3.06 m/s**
(b) $1.2 = v_2 \times A(0.02) \times 800$. $\therefore v_2 = $ **4.775 m/s**
(c) $p = \rho g h$. $\therefore (110 - 101.3) \times 10^3 = 800 \times 9.81 \times h$. $\therefore h = $ **1.109 m**

(d) $\dot{V} = \dfrac{\dot{m}}{\rho} = \dfrac{1.2}{800} = 0.0015 \ \text{m}^3/\text{s}$

For 6 m³/s, the time is $t = \dfrac{6}{0.0015} = 4000 \ \text{s} = $ **1.11 h**
(e) $V = Ah$, $\therefore 6 = A(1.5) \times h$. $\therefore h = $ **3.395 m**
(f) $p = \rho g h = 800 \times 9.81 \times 3.395 = $ **26.6 kPa (gauge) = 128 kPa (abs.)**

1.2 $Re = \dfrac{vd\rho}{\mu} = \dfrac{\text{m s}^{-1} \times \text{m} \times \text{kg m}^{-3}}{\text{Pas}} = \dfrac{\text{kg m}^{-1} \text{s}^{-1}}{\text{Pas}}$

Now $\text{Pa} = \text{N m}^{-2}$ and $\text{N} = \text{kg m s}^{-2}$. $\therefore \text{Pa} = \text{kg m}^{-1} \text{s}^{-2}$. $\therefore \text{Pas} = \text{kg m}^{-1} \text{s}^{-1}$,

$\therefore Re = \dfrac{vd}{v} = \dfrac{\text{m s}^{-1} \times \text{m}}{\text{m}^2 \text{ s}^{-1}} = $ **dimensionless**

$Re = \dfrac{\text{kg m}^{-1} \text{s}^{-1}}{\text{kg m}^{-1} \text{s}^{-1}} = $ **dimensionless**

1.3 Cross-sectional area $= 0.3 \times 0.2 = 0.06 \ \text{m}^2$
Wetted perimeter $= 2 \times 0.3 + 2 \times 0.2 = 1.0 \ \text{m}$

$d_e = 4 \times \dfrac{0.06}{1.0} = 0.24 \ \text{m}$

$Re = \dfrac{vd\rho}{\mu}$

$ = \dfrac{3 \times 0.24 \times 1.18}{18.5 \times 10^{-6}}$

$ = $ **45.9 × 10³**

1.4 $Re = \dfrac{vd\rho}{\mu}$

$2000 = \dfrac{0.5 \times d \times 998}{1.00 \times 10^{-3}}$

$\therefore d = $ **4.01 mm**

Chapter 2

2.1 (a) $v_2 = 3 \times \left(\dfrac{100}{80}\right)^2 = $ **4.69 m/s**

$$\frac{p_1}{\rho g} + \frac{v_1^2}{2g} + h_1 = \frac{p_2}{\rho g} + \frac{v_2^2}{2g} + h_2 + H_L$$

② is datum, so $h_2 = 0$. No head loss, $H_L = 0$

Substituting:

$$\frac{50 \times 10^3}{1.1 \times 10^3 \times 9.81} + \frac{3^2}{19.62} + 3 = \frac{p_2}{1.1 \times 10^3 \times 9.81} + \frac{4.69^2}{19.62}$$

$$4.6335 + 0.459 + 3 = \frac{p_2}{1.1 \times 10^3 \times 9.81} + 1.12$$

$$8.092 = \frac{p_2}{10.79 \times 10^3} + 1.12$$

$$\therefore p_2 = \textbf{75.23 kPa}$$

(b) v_2 is same $=$ **4.69 m/s**

If $H_L = 10\% \ H$, then $H_L = 0.1 \times 8.092 = 0.8092$ m

$$\therefore 8.092 = \frac{p_2}{10.79 \times 10^3} + 1.12 + 0.8092$$

$$\therefore p_2 = \textbf{66.5 kPa}$$

(c) $H_L = 0.8092$ m

$$\Delta p = \rho g H_L$$
$$= 1.1 \times 10^3 \times 9.81 \times 0.8092$$
$$= \textbf{8.73 kPa}$$

Also $\Delta p = 75.23 - 66.5 = $ **8.73 kPa**

2.2 $v = \dfrac{\dot{V}}{A} = \dfrac{1.5 \times 10^{-3}}{A(0.025)} = 3.056$ m/s

(a) $H_L = f\dfrac{L}{d}\dfrac{v^2}{2g} = 0.018 \times \dfrac{12}{0.025} \times \dfrac{3.056^2}{19.62} = $ **4.112 m**

(b) $\Delta p = \rho g H_L = 750 \times 9.81 \times 4.112 = $ **30.25 kPa**

(c) $P = \dot{m}gH = 1.5 \times 0.75 \times 9.81 \times 4.112 = $ **45.4 W**

2.3 $v = \dfrac{\dot{V}}{A} = \dfrac{20 \times 10^{-3}}{3600 \times A(0.01)} = 0.0707$ m/s

From Appendix 3, kerosene at 25°C, $\mu = 1.7 \times 10^{-3}$ Pas

$$Re = \frac{vd\rho}{\mu} = \frac{0.0707 \times 0.01 \times 780}{1.7 \times 10^{-3}} = 324.6 \ \therefore \text{ laminar}$$

$$f = \frac{64}{Re} = \frac{64}{324.6} = 0.1972$$

$$H_L = f\frac{L}{d}\frac{v^2}{2g} - 0.1972 \times \frac{1}{0.01} \times \frac{0.0707^2}{19.62} = 5.03 \times 10^{-3} \text{ m} = \textbf{5.03 mm}$$

2.4 (a) (i) $\varepsilon_R = 0.01$, $Re = 10^7$, diagram $\rightarrow f = $ **0.038**

$\varepsilon_R = 10 \times 10^{-3}$, $Re = 10 \times 10^6$

Using the formula:

$$f = 0.0055\left[1 + \left(20 \times 10 + \frac{1}{10}\right)^{\frac{1}{3}}\right] = \textbf{0.0377}$$

$$\% \text{ difference} = \frac{0.038 - 0.0377}{0.038} \times 100 = \textbf{0.87\%}$$

(ii) $\varepsilon_R = 0$ (smooth), $Re = 10^5$, diagram $\rightarrow f = \mathbf{0.0179}$
Using the formula with $Re = 0.1 \times 10^6$

$$f = 0.0055 \left[1 + \left(0 + \frac{1}{0.1} \right)^{\frac{1}{3}} \right] = \mathbf{0.0173}$$

$$\% \text{ difference} = \frac{0.0179 - 0.0173}{0.0179} \times 100 = \mathbf{3.1\%}$$

(iii) $\varepsilon_R = 0.002$, $Re = 10^4$, diagram $\rightarrow f = \mathbf{0.034}$
Using the formula with $\varepsilon_R = 2 \times 10^{-3}$, $Re = 0.01 \times 10^6$

$$f = 0.0055 \left[1 + \left(20 \times 2 + \frac{1}{0.01} \right)^{\frac{1}{3}} \right] = \mathbf{0.0341}$$

$$\% \text{ difference} = \frac{0.034 - 0.0341}{0.034} = \mathbf{-0.17\%}$$

(b) $\quad v = \dfrac{\dot{V}}{A} = \dfrac{0.4 \times 10^{-3}}{A(0.01)} = 5.093 \text{ m/s}$

$$Re = \frac{v d \rho}{\mu} = \frac{5.093 \times 0.01 \times 780}{1.7 \times 10^{-3}} = 23.37 \times 10^3$$

From Appendix 4, drawn tube, $\varepsilon = 0.0015$ mm

$$\varepsilon_R = \frac{0.0015}{10} = 0.00015$$

Using the formula, $f = 0.0055 \left[1 + \left(20\,000\varepsilon_R + \dfrac{10^6}{Re} \right)^{\frac{1}{3}} \right]$

$$= 0.0055 \left[1 + \left(3 + \frac{10^6}{23.37 \times 10^3} \right)^{\frac{1}{3}} \right]$$

$$= 0.0252$$

From the Moody diagram, $Re \approx 2.3 \times 10^4$, $\varepsilon_R = 0.00015$
$f = 0.0255$ (See Fig. S2.4 opposite.)
Using the value calculated by the formula:

$$H_L = f \frac{L}{d} \frac{v^2}{2g} = 0.0252 \times \frac{1}{0.01} \times \frac{5.093^2}{19.62} = \mathbf{3.33 \text{ m}}$$

2.5 (a) Denoting the surface of the open tank as ① and the surface of the pressurised tank as ②:

$$H = \frac{p_2 - p_1}{\rho g} + h_2 - h_1 + \frac{v_2^2 - v_1^2}{2g} + H_L$$

Now $\quad p_1 = 0$ (vented)
$\qquad h_1 = 0$ (datum)
$\qquad v_1 = 0$ (large tank)
$\qquad p_2 = 20$ kPa (gauge)
$\qquad v_2 = 0$ (large tank)
$\qquad h_2 = 4$ m

Substituting:

$$H = \frac{20 \times 10^3 - 0}{950 \times 9.81} + 4 + 0 + H_L$$

$$= 6.146 + H_L$$

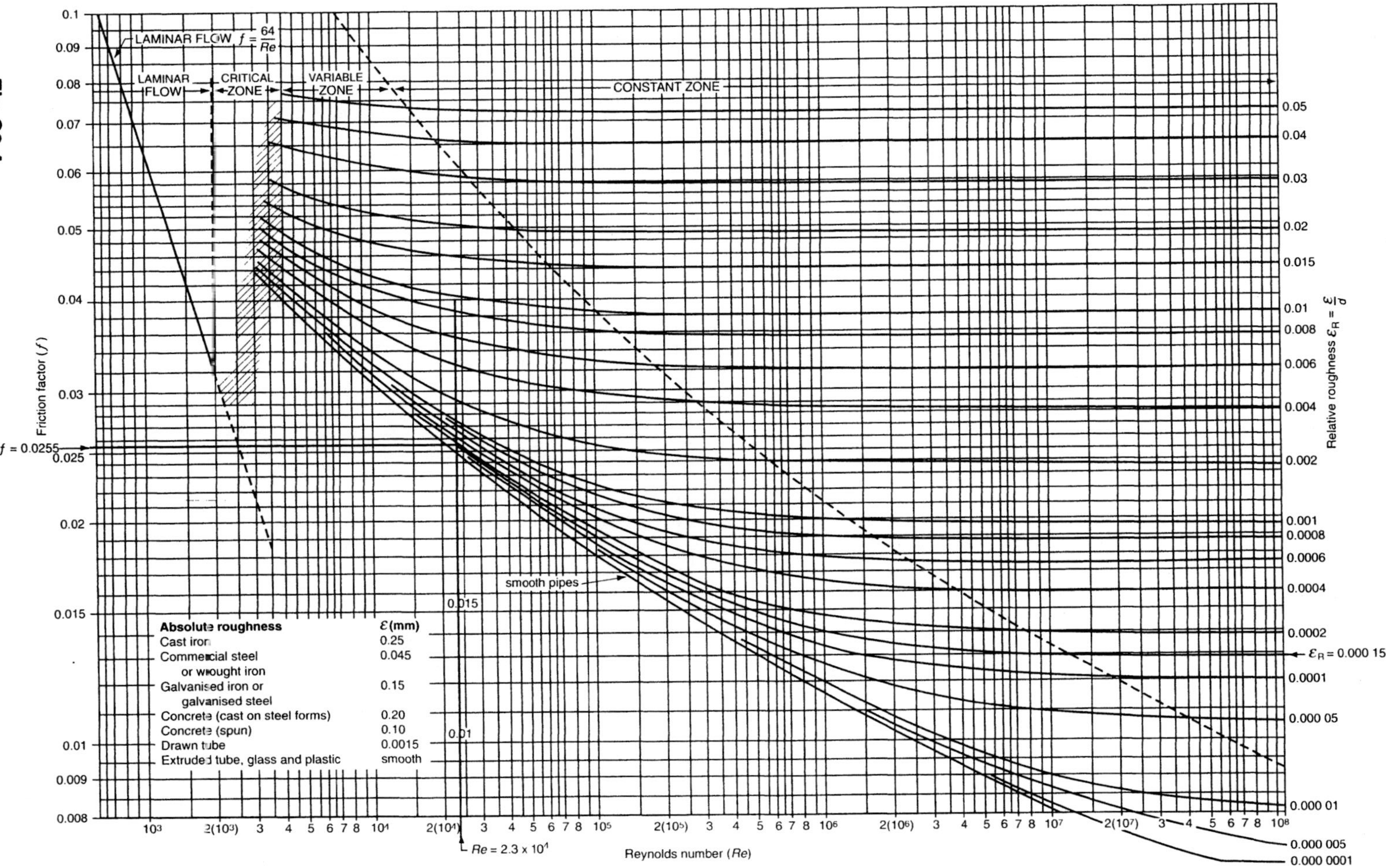
LAMINAR FLOW $f = \dfrac{64}{Re}$
LAMINAR FLOW
CRITICAL ZONE
VARIABLE ZONE
CONSTANT ZONE
Fig S2.4
Friction factor (f)
$f = 0.0255$
Relative roughness $\varepsilon_R = \dfrac{\varepsilon}{d}$
$\varepsilon_R = 0.000\,15$
smooth pipes
$Re = 2.3 \times 10^4$
Reynolds number (Re)
Absolute roughness
Cast iron
Commercial steel or wrought iron
Galvanised iron or galvanised steel
Concrete (cast on steel forms)
Concrete (spun)
Drawn tube
Extruded tube, glass and plastic
ε (mm)
0.25
0.045
0.15
0.20
0.10
0.0015
smooth

Now $\varepsilon = 0.0015$ mm (Appendix 4)

$$d = 30 \text{ mm}, \therefore \varepsilon_R = \frac{0.0015}{30} = 0.05 \times 10^{-3}$$

K factors from Appendix 6 are:

Sudden entrance   0.5
Sudden exit   1.0
2 gate values   0.4
1 globe value   6.0
       ΣK $\overline{7.9}$

$\upsilon = 2$ m/s

$$Re = \frac{\upsilon d \rho}{\mu} = \frac{2 \times 0.03 \times 950}{5 \times 10^{-3}} = 11.4 \times 10^3$$

$$f = 0.0055 \left[1 + \left(20\,000 \varepsilon_R + \frac{10^6}{Re} \right)^{\frac{1}{3}} \right]$$

$$= 0.0055 \times [1 + (1 + 87.7)^{\frac{1}{3}}]$$

$$= 0.03$$

$$H_L = \left(f \frac{L}{d} + \Sigma K \right) \frac{\upsilon^2}{2g}$$

$$= \left(0.03 \times \frac{6}{0.03} + 7.9 \right) \frac{\upsilon^2}{19.62}$$

$$= 0.7085 \upsilon^2$$

$\therefore$ the system head equation is:

$H = 6.146 + 0.7085\upsilon^2$

(b)

υ (m/s)	0	0.5	1.0	1.5	2.0	2.5	3.0
H (m)	6.15	6.32	6.85	7.74	8.98	10.6	12.5

These points may be plotted to obtain the system head curve as shown in Fig. S2.5 (opposite).

(c) $$\upsilon = \frac{10^{-3} \dot{V}}{A}, \text{ when } \dot{V} \text{ is in L/s}$$

$$d = 0.03 \text{ m}$$

$$\therefore \upsilon = \frac{10^{-3} \dot{V}}{A(0.03)} = 1.4147 \dot{V}$$

Substituting in the system head equation:

$$H = 6.146 + 0.7085 \times (1.4147 \dot{V})^2$$

$H = 6.146 + 1.418 \dot{V}^2$

Check at 2 L/s, $H = 6.146 + 1.418 \times 2^2$

$$\therefore \mathbf{H = 11.8 \text{ m}}$$

At 2 L/s, $\upsilon = \dfrac{0.002}{A(0.03)} = 2.83$ m/s

From graph Fig. S2.5, this checks.

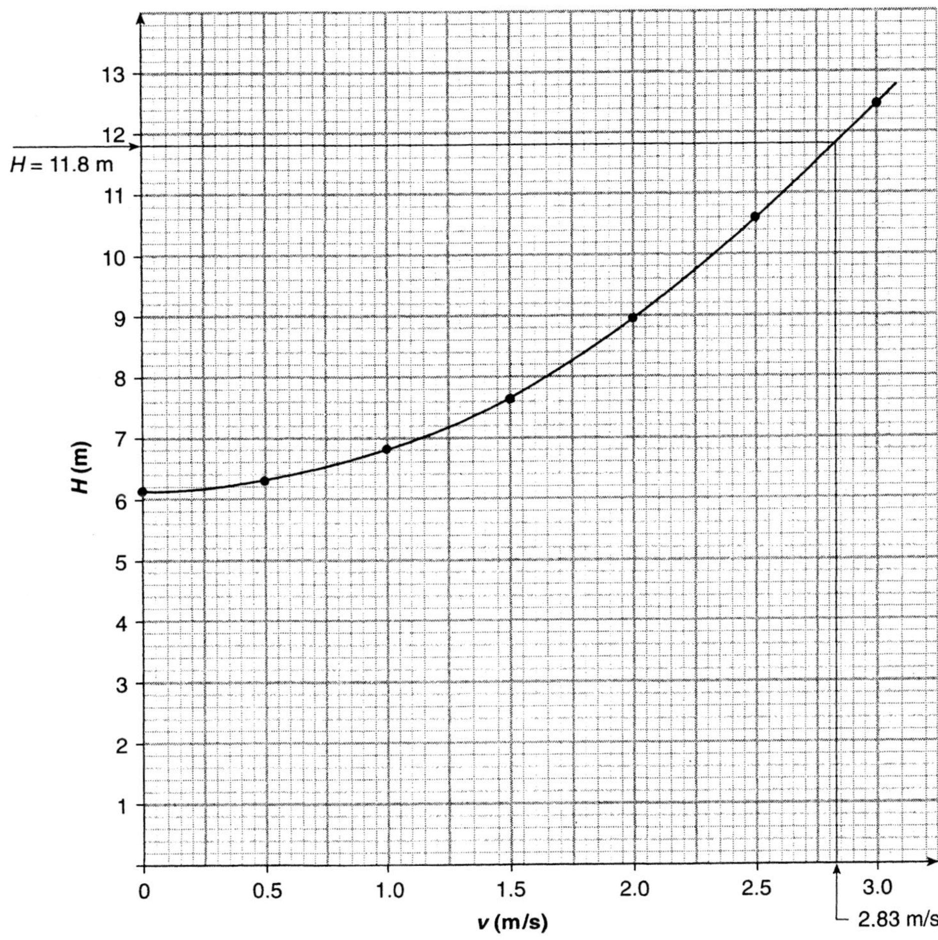

Fig. S2.5

Chapter 3

3.1 Denoting the surface of the pressurised tank as position (1) and the surface of the vented tank as position (2), and applying the Bernoulli equation between these positions:

$$\frac{p_1}{\rho g} + \frac{v_1^2}{2g} + h_1 - \frac{p_2}{\rho g} + \frac{v_2^2}{2g} + h_2 + H_{\mathrm{L}}$$

Now $p_1 = 20$ kPa, $v_1 = 0$ (large tank), $h_1 = 4$ m

$p_2 = 0$ (atmospheric), $v_2 = 0$ (large tank), $h_2 = 0$ (datum)

$$\therefore \quad \frac{20 \times 10^3}{950 \times 9.81} + 0 + 4 = 0 + 0 + 0 + H_{\mathrm{L}}$$

$$\therefore H_{\mathrm{L}} = 6.146 \text{ m}$$

Now $\varepsilon = 0.0015$ mm (Appendix 4), $d = 30$ mm, $\therefore \varepsilon_{\mathrm{R}} = \dfrac{0.0015}{30} = 0.05 \times 10^{-3}$

K factors from Appendix 6 are:

Sudden entrance	0.5
Sudden exit	1.0
2 gate valves	0.4
1 globe valve	6.0
ΣK	7.9

Trial 1, $v = 2$ m/s

$$Re = \frac{v d \rho}{\mu} = \frac{2 \times 0.03 \times 950}{5 \times 10^{-3}} = 11.4 \times 10^3$$

$$f = 0.0055\left[1 + \left(20000\varepsilon_R + \frac{10^6}{Re}\right)^{\frac{1}{3}}\right]$$

$$= 0.0055[1 + (1 + 87.7)^{\frac{1}{3}}]$$

$$= 0.03$$

$$H_L = \left(f\frac{L}{d} + \Sigma K\right)\frac{v^2}{2g} = \left(0.03 \times \frac{6}{0.03} + 7.9\right)\frac{v^2}{19.62} = 0.709v^2$$

$$\therefore 0.709v^2 = 6.146, \therefore v = 2.945 \text{ m/s}$$

Trial 2, $v = 2.945$ m/s

$Re = 16.8 \times 10^3$

$f = 0.0271$

$$H_L = \left(0.0271 \times \frac{6}{0.03} + 7.9\right)\frac{v^2}{19.62} = 0.679v^2$$

$$\therefore 0.679v^2 = 6.146, \therefore v = \textbf{3 m/s}$$

3.2 (a) Calling the 160 mm diameter pipe, A, and the 150 mm diameter pipe, B:

$$\frac{L_E}{d_E^5} = \frac{L_A}{d_A^5}, \text{ or } L_E = L_A\left(\frac{d_E}{d_A}\right)^5$$

$$\therefore L_E = 60 \times \left(\frac{150}{160}\right)^5 = 43.45 \text{ m}$$

$$\therefore L_E \text{ (total)} = 35 + 43.45 = \textbf{78.45 m}$$

(b) $\Sigma K = 0.5$ (entrance) $+ 1.0$ (exit) $+ 0.9$ (reducing elbow)

$$= 2.4$$

$$\varepsilon_R = \frac{0.045}{150} = 0.3 \times 10^{-3}$$

$$Re = \frac{vd}{v} = \frac{v \times 0.15}{1.1 \times 10^{-6}} = 0.1364 \times 10^{-6}v$$

$$f = 0.0055\left[1 + \left(20000\varepsilon_R + \frac{10^6}{Re}\right)^{\frac{1}{3}}\right]$$

$$= 0.0055\left[1 + \left(6 + \frac{7.33}{v}\right)^{\frac{1}{3}}\right] \qquad \ldots (1)$$

The Bernoulli equation reduces to: $h = H_L$

$$\therefore 50 = \left(f\frac{L}{d} + \Sigma K\right)\frac{v^2}{2g}$$

$$\therefore 50 = \left(f \times \frac{78.45}{0.15} + 2.4\right)\frac{v^2}{19.62}$$

$$\therefore 50 = (523f + 2.4)\frac{v^2}{19.62} \qquad \ldots (2)$$

Trial 1, $v = 10$ m/s
Substituting in (1):

$$f = 0.0055 \times \left[1 + \left(6 + \frac{7.33}{10}\right)^{\frac{1}{3}}\right] = 0.0159$$

Substituting in (2):

$$50 = (523 \times 0.0159 + 2.4) \times \frac{v^2}{19.62}$$

$$\therefore v = 9.57 \text{ m/s}$$

Trial 2, $v = 9.57$ m/s
From (1): $f = 0.0159$
Since f has not changed, the solution is $v = 9.57$ m/s
$\dot{V} = vA = 9.57 \times A(0.15) = 0.169$ m³/s = **169 L/s**

3.3 **(a)** From Appendix 4, $\varepsilon = 0.15$ mm

$$\therefore \varepsilon_R = \frac{0.15}{300} = 0.5 \times 10^{-3}$$

$$Re = \frac{vd}{v} = \frac{1.4147 \times 0.3}{1.1 \times 10^{-6}} = 0.386 \times 10^6$$

$$f = 0.0055\left[1 + \left(20\,000\varepsilon_R + \frac{10^6}{Re}\right)^{\frac{1}{3}}\right]$$

$$= 0.0055 \times [1 + (10 + 2.592)^{\frac{1}{3}}]$$
$$= \mathbf{0.0183}$$

(b) Since f is the same in all pipes, the flow rate in pipes B and C will not change.
From Example 3.8, $\dot{V}_B = $ **74.3 L/s**
$$\dot{V}_C = \textbf{25.7 L/s}$$

(c) From Example 3.9, $L_E = 676$ m

$$H_L = f\frac{L}{d}\frac{v^2}{2g} = 0.0183 \times \frac{676}{0.3} \times \frac{1.4147^2}{19.62}$$
$$= \mathbf{4.21 \text{ m}}$$

3.4 First calculate the equivalent lengths of pipes A and B. Because the friction factor is assumed to be the same, use Formula 3.5:

$$\left[\frac{d_E^{\,5}}{L_E}\right]^{\frac{1}{2}} = \left[\frac{d_A^{\,5}}{L_A}\right]^{\frac{1}{2}} + \left[\frac{d_B^{\,5}}{L_B}\right]^{\frac{1}{2}}$$

Now $d_E = d_B = d_A = d$ (say)
$L_E = ?$ $L_A = L_B = 6$ km

Substituting:

$$\left[\frac{d^5}{L_E}\right]^{\frac{1}{2}} = \left[\frac{d^5}{6}\right]^{\frac{1}{2}} + \left[\frac{d^5}{6}\right]^{\frac{1}{2}} = 2\left[\frac{d^5}{6}\right]^{\frac{1}{2}}$$

$$\therefore \frac{d^5}{L_E} = 4\,\frac{d^5}{6}$$

$$\therefore L_E = \frac{6}{4} = 1.5 \text{ km}$$

$$\therefore L_E \text{ (total)} = 1.5 + 4 = 5.5 \text{ km}$$

$$\dot{V} = vA, \therefore 0.5 = \frac{\pi d^2}{4} \times v, \therefore d = \frac{0.7979}{\sqrt{v}}$$

Trial 1, $v = 1$ m/s, $\therefore d = 0.7979$ m, say 0.8 m

$$\varepsilon_R = \frac{0.2}{800} = 0.25 \times 10^{-3}$$

$$Re = \frac{vd}{\nu} = \frac{1 \times 0.8}{1.15 \times 10^{-6}} = 0.696 \times 10^6$$

$$f = 0.0055\left[1 + \left(20\,000\,\varepsilon_R + \frac{10^6}{Re}\right)^{\frac{1}{3}}\right]$$

$$= 0.0055[1 + (5 + 1.438)^{\frac{1}{3}}]$$
$$= 0.0157$$

Applying the Bernoulli equation: $h = H_L$

$$\therefore 20 = f\frac{L}{d}\frac{v^2}{2g}$$

$$\therefore 20 = 0.0157 \times \frac{5500}{0.8} \times \frac{v^2}{19.62}$$

$$\therefore v = 1.9 \text{ m/s}$$

Trial 2, $v = 1.9$ m/s

$$d = \frac{0.7979}{\sqrt{1.9}} = 0.58 \text{ m}$$

$$\varepsilon_R = 0.345 \times 10^{-3}$$
$$Re = 0.958 \times 10^6$$
$$f = 0.0166$$

$$20 = 0.0166 \times \frac{5500}{0.58} \times \frac{v^2}{19.62}$$

$$\therefore v = 1.58 \text{ m/s}$$

Trial 3, $v = 1.6$ m/s

$$d = \frac{0.7979}{\sqrt{1.6}} = 0.63 \text{ m}$$

$$\varepsilon_R = 0.318 \times 10^{-3}$$
$$Re = 0.877 \times 10^6$$
$$f = 0.0163$$

$$20 = 0.0163 \times \frac{5500}{0.63} \times \frac{v^2}{19.62}$$

$$\therefore v = 1.66 \text{ m/s}$$

Trial 4, $v = 1.66$ m/s

$$d = 0.62 \text{ m}$$
$$\varepsilon_R = 0.323 \times 10^{-3}$$
$$Re = 0.895 \times 10^6$$
$$f = 0.0163$$

$$20 = 0.0163 \times \frac{5500}{0.62} \times \frac{v^2}{19.62}$$

$$\therefore v = 1.65 \text{ m/s}$$

$$\therefore d = \frac{0.7979}{\sqrt{1.65}} = 0.622 \text{ m}$$

Solution is $d = 0.622$ m, that is **622 mm**

Chapter 4

4.1 When the pipe is half-full:

$$A = 0.5 \times A(200) = 15\,708 \text{ mm}^2$$

Perimeter $P = 0.5 \times \pi \times 200 = 314.16$ mm

$$d_e = \frac{4A}{P} = \frac{4 \times 15\,708}{314.16} = 200 \text{ mm (same as the actual diameter)}$$

$$\varepsilon_R = \frac{\varepsilon}{d_e} = \frac{0.045}{200} = 0.225 \times 10^{-3}$$

$$Re = \frac{vd_e\rho}{\mu} = \frac{1 \times 0.2 \times 800}{2 \times 10^{-3}} = 0.08 \times 10^6$$

$$f = 0.0055\left[1 + \left(20\,000\varepsilon_R + \frac{10^6}{Re}\right)^{\frac{1}{3}}\right]$$

$$= 0.0055 \times \left[1 + \left(20 \times 0.225 + \frac{1}{0.08}\right)^{\frac{1}{3}}\right]$$

$$= 0.0196$$

Applying Formula 4.1:

$$v = \sqrt{\frac{Sd_e\,2g}{f}}$$

$$\therefore 1 = \sqrt{\frac{S \times 0.2 \times 19.62}{0.0196}}$$

$$\therefore 1 = \sqrt{199.8S}$$

$$\therefore S = \mathbf{1/200} \text{ (slope of 1 in 200)}$$

4.2 (a)

$$A = h^2, \quad P = \frac{2h}{\cos 45°} = 2.828h$$

$$R = \frac{A}{P} = \frac{h^2}{2.828h} = 0.3536h$$

$$\dot{V} = vA, \quad \therefore 5 = v \times h^2, \quad \therefore v = \frac{5}{h^2}$$

$$v = C\sqrt{RS}$$

$$\therefore \frac{5}{h^2} = 60\sqrt{0.3536h \times \frac{1}{900}}$$

Squaring both sides:

$$\frac{25}{h^4} = 3600 \times 0.3928 \times 10^{-3}h$$

$$= 1.4142h$$

$$\therefore 17.677 = h^5$$

$$\therefore h = 1.776 \text{ m} = \textbf{1.78 m} \text{ (to 3 significant figures)}$$

(b) $\quad v = \dfrac{5}{h^2} = \dfrac{5}{1.776^2} = \textbf{1.585 m/s}$

(c) From Formula 4.4:

$$C = \sqrt{\frac{8g}{f}}$$

$$\therefore 60 = \sqrt{\frac{8 \times 9.81}{f}}$$

$$\therefore f = \textbf{0.0218}$$

(d) $\quad R = 0.3536h = 0.3536 \times 1.776 = 0.628 \text{ m}$

$$v = C\sqrt{RS}$$

$$= 60\sqrt{0.628 \times \frac{1}{900}}$$

$$= \textbf{1.585 m/s} \quad \text{(Checks)}$$

4.3 **(a)** $A = 3 \times 1 = 3 \text{ m}^2$

$$P = 3 + 1 + 1 = 5 \text{ m}$$

$$R = \frac{A}{P} = \frac{3}{5} = 0.6 \text{ m}$$

$$v = \frac{\dot{V}}{A} = \frac{5}{3} \text{ m/s}$$

From Appendix 9, for a smooth cement channel, $n = 0.012 \text{ m}^{-\frac{1}{3}}\text{ s}$
Using the Manning formula:

$$v = \frac{1}{n} R^{\frac{2}{3}} S^{\frac{1}{2}}$$

$$\therefore \frac{5}{3} = \frac{1}{0.012} \times 0.6^{\frac{2}{3}} \times S^{\frac{1}{2}}$$

$$\therefore S^{\frac{1}{2}} = 0.0281$$

$$\therefore S = \textbf{0.79} \times \textbf{10}^{-3} \text{ or } \textbf{1 in 1265}$$

(b) $\quad C = \dfrac{R^{\frac{1}{6}}}{n} = \dfrac{0.6^{\frac{1}{6}}}{0.012} = \textbf{76.5 m}^{\frac{1}{2}}\textbf{ s}^{-1}$

Chapter 5

5.1 **(a)** $F = pA = 2.5 \times A(50) = 4909 \text{ N} = \textbf{4.91 kN}$

(b) $F = 2.5 \times [A(50) - A(25)] = 3681 \text{ N} = \textbf{3.68 kN}$

(c) $\dot{V} = A(0.05) \times 0.2 \text{ m}^3/\text{s} = \textbf{0.393 L/s}$

(d) $P = Fv = 4909 \times 0.2 = \textbf{982 W}$

(e) $P_f = p\dot{V} = 2.5 \times 10^6 \times 0.393 \times 10^{-3} = \textbf{982 W}$

5.2 Calculate the specific speed in each case:

(a) $N_s = \dfrac{N\sqrt{\dot{V}}}{H^{\frac{3}{4}}} = \dfrac{1450 \times \sqrt{30}}{2^{\frac{3}{4}}} = 4722$

(b) $N_s = \dfrac{N\sqrt{\dot{V}}}{H^{\frac{3}{4}}} = \dfrac{1450 \times \sqrt{20}}{10^{\frac{3}{4}}} = 1153$

(c) $N_s = \dfrac{N\sqrt{\dot{V}}}{H^{\frac{3}{4}}} = \dfrac{1450 \times \sqrt{10}}{15^{\frac{3}{4}}} = 602$

(d) $N_s = \dfrac{N\sqrt{\dot{V}}}{H^{\frac{3}{4}}} = \dfrac{1450 \times \sqrt{6}}{40^{\frac{3}{4}}} = 223$

From Figure 5.23, the following selections may be made:
(a) **Mixed-flow pump, divergent-cone type.**
(b) **Centrifugal pump, low *OD/ID* ratio.**
(c) **Centrifugal pump, medium *OD/ID* ratio.**
(d) **Positive-displacement pump or multistage centrifugal pump.**

5.3 (a) $P = T\omega$

$\therefore 11\,500 = T \times \pi \times \dfrac{2950}{30}$

$\therefore T = \textbf{37.2 Nm}$

(b) $\eta = \dfrac{\dot{m}gH}{P} = \dfrac{25 \times 9.81 \times 35}{11500} = 0.746 = \textbf{74.6\%}$

(c) $P_i = \dfrac{11.5}{0.84} = 13.7\text{ kW}$

Cost per day $= 13.7 \times 0.11 \times 8 = \textbf{\$12.05}$

5.4 The fluid power at 1440 rpm is:
$P_f = p\dot{V} = 530 \times 10^3 \times 0.6 \times 10^{-3} = 318\text{ W}$
The efficiency at 1440 rpm is:

$\eta = \dfrac{P_f}{P} = \dfrac{318}{400} = 0.795$

(a) At 2950 rpm, the volume flow rate is:

$\dot{V} = \dfrac{2950}{1440} \times 0.6 = 1.229\text{ L/s} = \textbf{1.3 L/s}$

(b) At 2950 rpm, the fluid power is:

$P_f = p\dot{V} = 530 \times 10^3 \times 1.229 \times 10^{-3} = 651.5\text{ W}$

The efficiency at 2950 rpm is the same:

$\eta = \dfrac{P_f}{P}, \therefore \dfrac{651.5}{P} = 0.795$

$\therefore P = \textbf{819 W}$

Alternatively: $P = \dfrac{400 \times 2950}{1440} = \textbf{819 W}$

5.5 **(a)** At 20 L/s, the following values can be read off the curves:

Impeller size (mm)	Head (m)	Power (kW)	Efficiency (%)	NPSHR (m)
260	22.4	5.75	77.2	1.0
230	16.25	4.2	76.8	1.0
204	11.5	3.0	75.0	1.0

(b) *260 mm impeller*

$$P_f = \dot{m}gH = 20 \times 9.81 \times 22.4 = \textbf{4.395 kW}$$

$$\eta = \frac{P_f}{P} = \frac{4.395}{5.75} = 0.764 = \textbf{76.4\%}$$

230 mm impeller

$$P_f = \dot{m}gH = 20 \times 9.81 \times 16.25 = \textbf{3.188 kW}$$

$$\eta = \frac{P_f}{P} = \frac{3.188}{4.2} = 0.759 = \textbf{75.9\%}$$

260 mm impeller

$$P_f = \dot{m}gH = 20 \times 9.81 \times 11.5 = \textbf{2.256 kW}$$

$$\eta = \frac{P_f}{P} = \frac{2.256}{3.0} = 0.752 = \textbf{75.2\%}$$

Note   The difference between the calculated values and the values read from the performance curves is due to the difficulty of plotting and reading charts exactly.

5.6 **(a)** $\eta = \dfrac{\dot{m}gH}{P}$

Hence the efficiency can be calculated as shown in the table:

Flow rate (L/s)	0	10	20	30
Head (m)	23	22	20.5	15.5
Power (kW)	3.8	4.3	5.7	6.5
Efficiency (%)	0	50.2	70.6	70.2

The performance curves may be drawn as the full lines in Fig. S5.6 (opposite).

(b) Using the affinity relations, the following table can be drawn up:

Original flow rate (L/s)	0	10	20	30
New flow rate (L/s)	0	6.69	13.38	20.07
New head (m)	10.29	9.85	9.175	6.938
New power (kW)	1.138	1.288	1.707	1.946
New efficiency (%)	0	50.2	70.6	70.2

Note that the efficiency is the same.

The performance curves may be drawn as the broken lines in Fig. S5.6 (opposite).

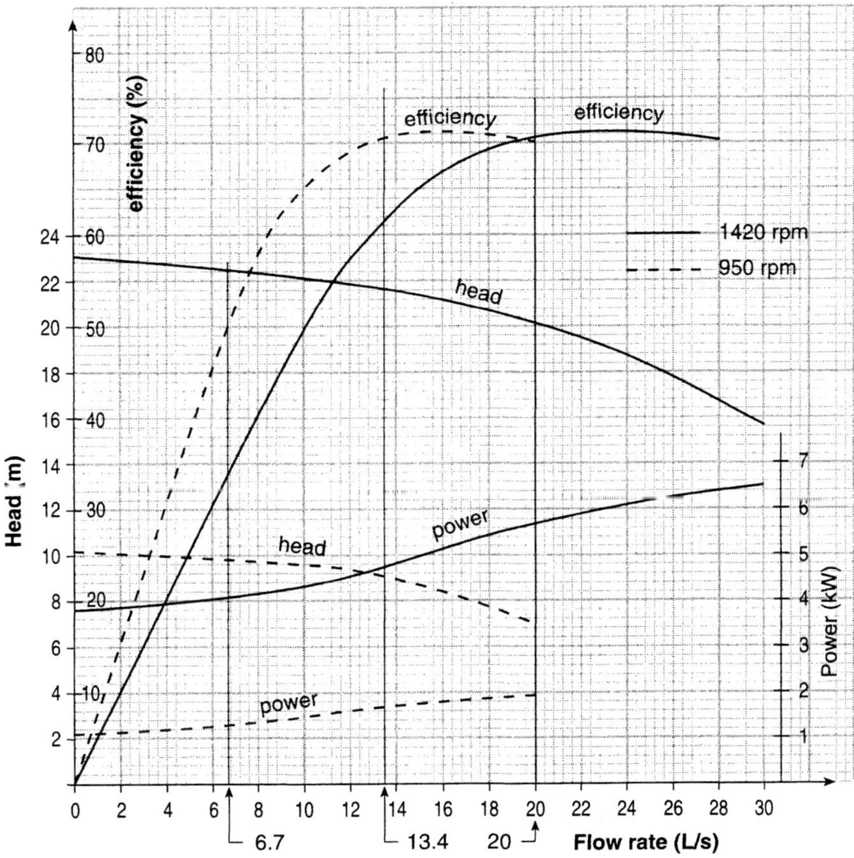

Fig. S5.6

(c) Specific speed $N_s = \dfrac{N\sqrt{\bar{V}}}{H^{\frac{3}{4}}}$

$$N_{s_1} = \frac{1420 \times \sqrt{20}}{20.5^{\frac{3}{4}}} = 659$$

$$N_{s_2} = \frac{950 \times \sqrt{13.38}}{9.175^{\frac{3}{4}}} = 659$$

Note that the specific speed is the same.

Chapter 6

6.1 (a) Calculating the velocity at each flow rate and using the system head equation, the following table can be derived:

Flow rate (L/s)	0	4	8	12	16
Velocity (m/s)	0	0.796	1.5915	2.387	3.183
System head (m)	4.5	4.75	5.49	6.72	8.45

Alternatively the system head equation can be written in terms of flow rate as follows:

$$v = \frac{\dot{V} \times 10^{-3}}{A(0.08)} \text{ where } \dot{V} \text{ is in L/s,}$$

$$\therefore v = 0.199\,\dot{V}$$

Substituting in the system head equation:

$$H = 4.5 + 0.39 \times (0.199\,\dot{V})^2$$

$$\therefore H = 4.5 + 0.0154\,\dot{V}^2$$

By substituting in this equation, the system head can be determined directly from the flow rate in L/s (without calculating the velocity). For example, at 16 L/s:

$$H = 4.5 + 0.0154 \times 16^2 = 8.45 \text{ m (as above)}.$$

The system head curve can then be plotted on the pump performance curve as shown in Figure S6.1:

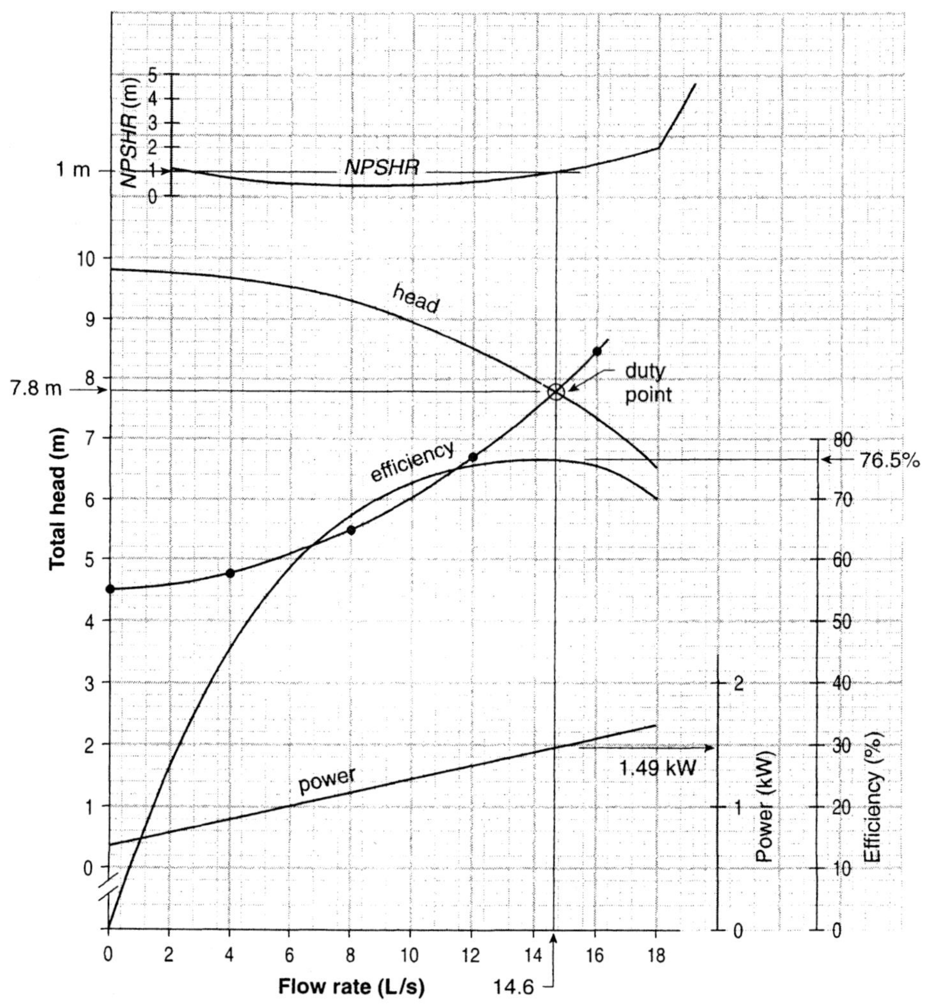

Fig. S6.1

The duty point can be located and the following values read off:
$\dot{V}$ = **14.6 L/s**, H = **7.8 m**, **P** = **1.49 kW**, η = **76.5%**, *NPSHR* = **1.0 m**

(b) The fluid power is given by $P_f = \dot{m}gh$

At the duty point, P_f = 14.6 × 9.81 × 7.8 = **1117 W**

The efficiency is $\eta = \dfrac{P_f}{P} = \dfrac{1117}{1490}$ = **75%**   (Close)

6.2 (a) The dynamic head is: $H_{\mathrm{dyn}} = \dfrac{v_2^2 - v_1^2}{2g} + \left(f\dfrac{L}{d} + \Sigma K\right)\dfrac{v^2}{2g}$

Now v_1 = 0 (pond), v_2 = 0 (large tank) ,
f = 0.015 (given), L = 20 m, ΣK = 9.5 (given)
Substituting:

$$H_{\mathrm{dyn}} = 0 + \left(\dfrac{0.015 \times 20}{0.08} + 9.5\right)\dfrac{v^2}{19.62}$$
$$= 0.6753v^2$$

The static head is :

$$H_{\mathrm{stat}} = \dfrac{p_2 - p_1}{\rho g} + h_2 - h_1$$

Now p_2 = 0, p_1 = 0, and $h_2 - h_1$ = 3 m
$\therefore H_{\mathrm{stat}}$ = 3 m
Therefore the system head equation (in terms of velocity) is:
H = 3 + 0.6753v^2
The system head may now be calculated at various flow rates (in the range 0–16 L/s).
The values thus obtained are given in the table below:

$\dot{V}$ (L/s)	0	4	8	12	16
v (m/s)	0	0.796	1.592	2.387	3.183
H (m)	3	3.43	4.71	6.85	9.84

These points can now be plotted on the pump performance curve as shown in Figure S6.2 on page 200.
The following values can be read off at the duty point:
$\dot{V}$ = **13.6 L/s**, **P** = **1.45 kW**, η = **76.5%**

(b) From Appendix 6, for a half-open globe valve, K = 12.0, whereas for a fully open one, K = 6.0.
$\therefore$ the new value of ΣK is given by:
ΣK = 9.5 + 6.0 = 15.5
Substituting:

$$H_{\mathrm{dyn}} = 0 + \left(\dfrac{0.015 \times 20}{0.08} + 15.5\right)\dfrac{v^2}{19.62}$$
$$= 0.981v^2$$
H_{stat} = 3 m

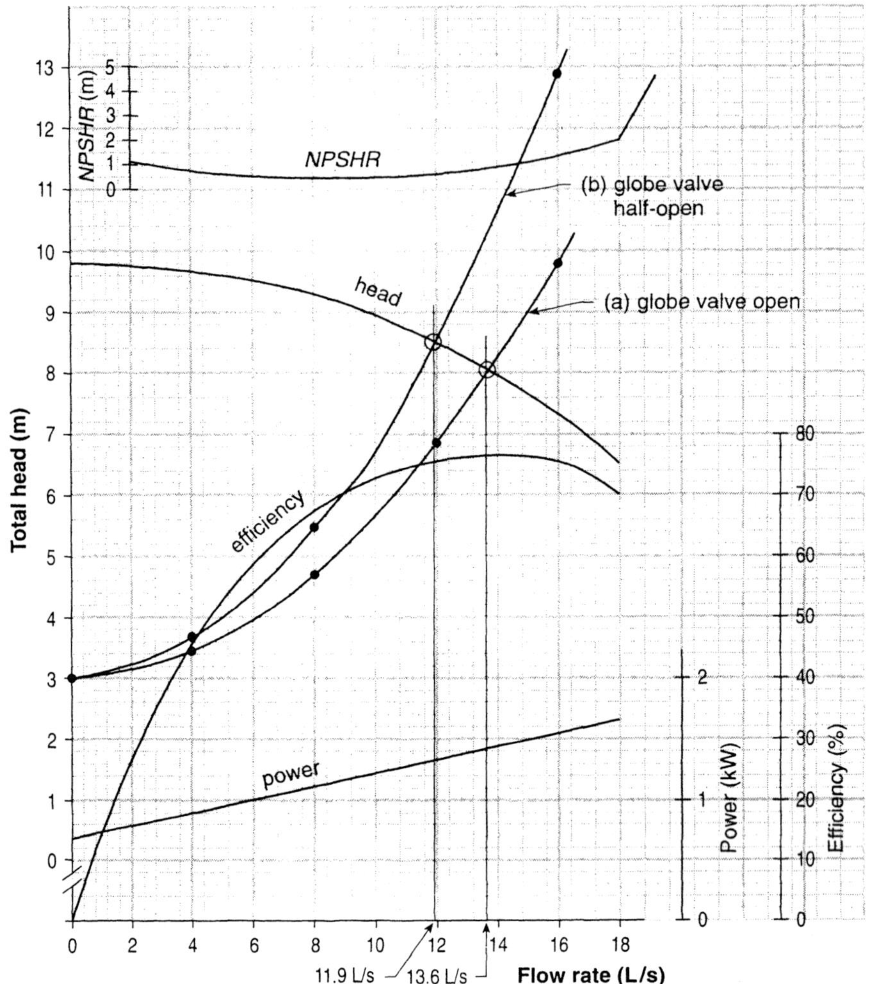

Fig. S6.2

Therefore the system head equation (in terms of velocity) is:

$$H = 3 + 0.981\upsilon^2$$

The system head can now be recalculated as shown in the table:

$\dot{V}$ (L/s)	0	4	8	12	16
υ (m/s)	0	0.796	1.592	2.387	3.183
H (m)	3	3.62	5.49	8.59	12.9

These points can now be plotted on the pump performance curve (refer to Figure S6.2).

The following new values can be read off at the new duty point:

$\dot{V}$ = **11.9 L/s**, P = **1.35 kW**, η = **75.6%**

6.3 **(a)** Applying the Bernoulli equation between the sump and the pump inlet:

$$\frac{p_i}{\rho g} = \frac{p_1}{\rho g} - \frac{v^2}{2g} - h - H_L$$

Now $v = \dfrac{0.035}{A(0.15)} = 1.981$ m/s

$$H_L = \left(f\frac{L}{d} + \Sigma K\right)\frac{v^2}{2g}$$

$$= \left(0.025 \times \frac{3}{0.15} + 15.5\right) \times \frac{1.981^2}{19.62} = 3.2 \text{ m}$$

Given $h = 2$ m, and $p_1 = 101.3$ kPa (abs.), and from Appendix 13, for water at $40°C$, $\rho = 992$ kg/m^3

Substituting:

$$\frac{p_i}{\rho g} = \frac{101.3 \times 10^3}{992 \times 9.81} - \frac{1.981^2}{19.62} - 2 - 3.2$$

$$= 5.01 \text{ m}$$

$$\therefore p_i = 5.01 \times 992 \times 9.81 \text{ Pa} = \textbf{18.75 kPa (abs.)}$$

(b) Using Formula 5.2:

$$NPSHA = \frac{p_i - p_v}{\rho g} + \frac{v^2}{2g}$$

From Appendix 13, for water at $40°C$, $p_v = 7.38$ kPa (abs.)

Substituting:

$$NPSHA = \frac{48.75 - 7.38}{0.992 \times 9.81} + \frac{1.981^2}{19.62} = \textbf{4.45 m}$$

(c) Using Formula 6.1:

$$NPSHA = \frac{p_1 - p_v}{\rho g} - h - H_L$$

$$= \frac{101.3 - 7.38}{0.992 \times 9.81} - 2 - 3.2$$

$$= 9.651 - 5.2$$

$$= \textbf{4.45 m} \text{ (as derived by Formula 5.2)}$$

(d) The difference between $NPSHA$ and $NPSHR = 4.45 - 2.2 = 2.25$ m
Maximum elevation occurs when $NPSHA = NPSHR$
Assuming the same head loss, the maximum elevation $h = 2 + 2.25 = \textbf{4.25 m}$

(e) If the elevation of the pump is h, then the length of the suction line is $h + 1$

$$H_L = \left(f\frac{L}{d} + \Sigma K\right)\frac{v^2}{2g}$$

$$= \left[0.025 \times \frac{(h + 1)}{0.15} + 15.5\right] \times \frac{1.981^2}{19.62}$$

$$= \left[0.025 \times \frac{(h + 1)}{0.15} + 15.5\right] \times 0.2$$

$$= 0.0333(h + 1) + 3.1$$

$$= 0.0333h + 3.133$$

Using Formula 6.1:

$$NPSHA = \frac{p_1 - p_v}{\rho g} - h - H_L$$

The value of $\dfrac{p_1 - p_v}{\rho g}$ as found in (c) = 9.651 does not change.

$\therefore NPSHA = 9.651 - h - H_L$

Substituting for H_L:

$$NPSHA = 9.651 - h - 0.0333h - 3.133$$
$$= 9.651 - 1.0333h - 3.133$$
$$= 6.518 - 1.0333h$$

Maximum elevation occurs when $NPSHA = NPSHR$,

$\therefore 2.2 = 6.518 - 1.0333h$

$\therefore h = \mathbf{4.18\ m}$

6.4 (a) $\dfrac{0.5 \times 10^{-3}}{6 \times 0.9} = \dfrac{\pi d^2}{4} \times d \times \dfrac{960}{60}$

$\therefore d^3 = 7.368 \times 10^{-6}$

$\therefore d = 0.0195\ \text{m} = \mathbf{19.5\ mm}$

(b) $P_f = p\dot{V} = 4 \times 10^6 \times 0.5 \times 10^{-3}\ \text{W} = 2\ \text{kW}$

$\eta = \dfrac{P_f}{P}$

$\therefore 0.8 = \dfrac{2}{P}$

$\therefore P = \mathbf{2.5\ kW}$

Appendixes

Appendix 1   List of principal symbols

Symbol	Meaning	Base units or value
A	area of a circle	m^2
B	factor (constant in the system head equation)	—
C	Chezy coefficient	$m^{\frac{1}{2}}\ s^{-1}$
	factor (constant in the system head equation)	—
d	diameter of a circle	m
d_e	equivalent diameter	m
f	friction factor	—
g	gravitational constant	9.81 m^2/s (or N/kg)
h	vertical height	m
	hoad of a fluid	m
H	total head of a fluid	m
	system head	m
H_L	head loss	m
K	K factor for a fitting	—
ΣK	sum of the K factors for a number of fittings in series	—
L_E	equivalent length	m
m	mass	kg
$\dot{m}$	mass flow rate	kg/s
n	Manning coefficient	$m^{-\frac{1}{3}}\ s$
N	rotational speed of a shaft	rpm
N_s	specific speed	various*
$NPSHA$	net positive suction head available	m
$NPSHR$	net positive suction head required	m
p	pressure	Pa
p_i	pressure at the inlet of a pump	Pa
p_v	saturation vapour pressure of a fluid	Pa
P	perimeter of a circle	m
	shaft power	W
P_f	fluid power	W
R	hydraulic radius	m
Re	Reynolds number	—
RD	relative density	—
S	slope of a channel	—
t	time	s

---

* units of specific speed depend on the units used for rotational speed, flow rate, head and power, and are not usually quoted.

Symbol	Meaning	Base units or value
T	torque	Nm
v	specific volume	m^3/kg
V	volume	m^3
$\dot{V}$	volume flow rate	m^3/s

Greek symbols

Symbol	Meaning	Base units or value
ε	absolute roughness	m
θ	angular displacement	rad
ω	angular velocity	rad/s
ρ	density	kg/m^3
μ	dynamic viscosity	Pas
η	efficiency	—
ν	kinematic viscosity	m^2/s
ε_R	relative roughness	—
υ	velocity	m/s

Appendix 2   List of principal formulas

$$Re = \frac{vd}{v} = \frac{vd\rho}{\mu}$$

Reynolds number (1.1)

$$d_e = \frac{4A}{P}$$

Equivalent diameter (1.2)

$$\frac{p_1}{\rho g} + \frac{v_1^2}{2g} + h_1 = \frac{p_2}{\rho g} + \frac{v_2^2}{2g} + h_2 + H_L$$

Bernoulli equation

$$H_L = f\frac{L}{d}\frac{v^2}{2g}$$

Head loss in a pipe (Darcy formula) (2.1)

$$f = \frac{64}{Re}$$

Friction factor—laminar flow (2.2)

$$\varepsilon_R = \frac{\varepsilon}{d}$$

Relative roughness (2.3)

$$f = 0.0055\left[1 + \left(20\,000\varepsilon_R + \frac{10^6}{Re}\right)^{\frac{1}{3}}\right]$$

Friction factor—turbulent flow (2.4)

$$H_L = K\frac{v^2}{2g}$$

Head loss in a fitting (2.5)

$$H_L = \Sigma K\frac{v^2}{2g}$$

Head loss in a number of fittings (2.6)

$$H_L = \left(f\frac{L}{d} + \Sigma K\right)\frac{v^2}{2g}$$

Head loss in a fluid system (2.7)

$$L_E = \frac{Kd}{f}$$

Equivalent length of a fitting (2.8)

$$H = A + Bv^2$$

System head equation—in terms of velocity (2.8)

$$H = A + C\dot{V}^2$$

System head equation—in terms of flow rate (2.9)

$$\frac{f_E L_E}{d_E^5} = \frac{f_A L_A}{d_A^5} + \frac{f_B L_B}{d_B^5} + \dots$$

Equivalent length series pipes (3.1)

$$\frac{L_E}{d_E^5} = \frac{L_A}{d_A^5} + \frac{L_B}{d_B^5} + \dots$$

Equivalent length series pipes (same friction factor) (3.2)

$$\frac{L_A v_A^2}{d_A} = \frac{L_B v_B^2}{d_B}$$

Parallel pipes (same friction factor) (3.3)

$$\left[\frac{d_E^5}{f_E L_E}\right]^{\frac{1}{2}} = \left[\frac{d_A^5}{f_A L_A}\right]^{\frac{1}{2}} + \left[\frac{d_B^5}{f_B L_B}\right]^{\frac{1}{2}} + \ \cdots\cdots\cdots \qquad \text{Equivalent length parallel pipes (3.4)}$$

$$\left[\frac{d_E^5}{L_E}\right]^{\frac{1}{2}} = \left[\frac{d_A^5}{L_A}\right]^{\frac{1}{2}} + \left[\frac{d_B^5}{L_B}\right]^{\frac{1}{2}} + \ \cdots\cdots\cdots \qquad \begin{array}{r}\text{Equivalent length parallel pipes}\\ \text{(same friction factor) (3.5)}\end{array}$$

$$v = \sqrt{\frac{S d_e\, 2g}{f}} \qquad\qquad\qquad \text{Darcy formula applied to channel flow (4.1)}$$

$$v = C\sqrt{RS} \qquad\qquad\qquad \text{Chezy formula (4.2)}$$

$$R = \frac{A}{P} = \frac{d_e}{4} \qquad\qquad\qquad \text{Hydraulic radius (4.3)}$$

$$C = \sqrt{\frac{8g}{f}} \qquad\qquad\qquad \text{Chezy coefficient (4.4)}$$

$$C = \frac{R^{\frac{1}{6}}}{n} \qquad\qquad\qquad \text{Chezy coefficient (4.5)}$$

$$v = \frac{1}{n} R^{\frac{2}{3}} S^{\frac{1}{2}} \qquad\qquad\qquad \text{Manning formula (4.6)}$$

$$n = R^{\frac{1}{6}}\sqrt{\frac{f}{8g}} \qquad\qquad\qquad \text{Manning coefficient (4.7)}$$

$$N_s = \frac{N\sqrt{\dot{V}}}{H^{\frac{3}{4}}} \qquad\qquad\qquad \text{Specific speed of a pump (5.1)}$$

$$NPSHA = \frac{p_i - p_v}{\rho g} + \frac{v^2}{2g} \qquad\qquad \text{Net positive suction head available (5.2)}$$

$$\eta = \frac{P_f}{P} = \frac{\dot{m} g H}{P} \qquad\qquad\qquad \text{Pump efficiency (5.3)}$$

$$\frac{\dot{V}_2}{\dot{V}_1} = \frac{N_2}{N_1} \qquad\qquad \text{Volume–speed relationship for a rotodynamic pump (5.4)}$$

$$\frac{H_2}{H_1} = \left[\frac{N_2}{N_1}\right]^2 \qquad\qquad \text{Head–speed relationship for a rotodynamic pump (5.5)}$$

$$\frac{P_2}{P_1} = \left[\frac{N_2}{N_1}\right]^3 \qquad\qquad \text{Power–speed relationship for a rotodynamic pump (5.6)}$$

$$\eta = \frac{P}{P_{\mathrm{f}}}$$

Motor or turbine efficiency (5.7)

$$N_{\mathrm{s}} = \frac{N\sqrt{P}}{H^{\frac{5}{4}}}$$

Specific speed of a motor or turbine (5.8)

$$NPSHA = \frac{p_1 - p_{\mathrm{v}}}{\rho g} - h - H_{\mathrm{L}}$$

Net positive suction head available (6.1)

$$\eta = \frac{P}{P_{\mathrm{f}}}$$

$$N_{\mathrm{s}} = \frac{N\sqrt{P}}{H^{\frac{5}{4}}}$$

Appendix 3 Viscosity variation with temperature, for some common fluids

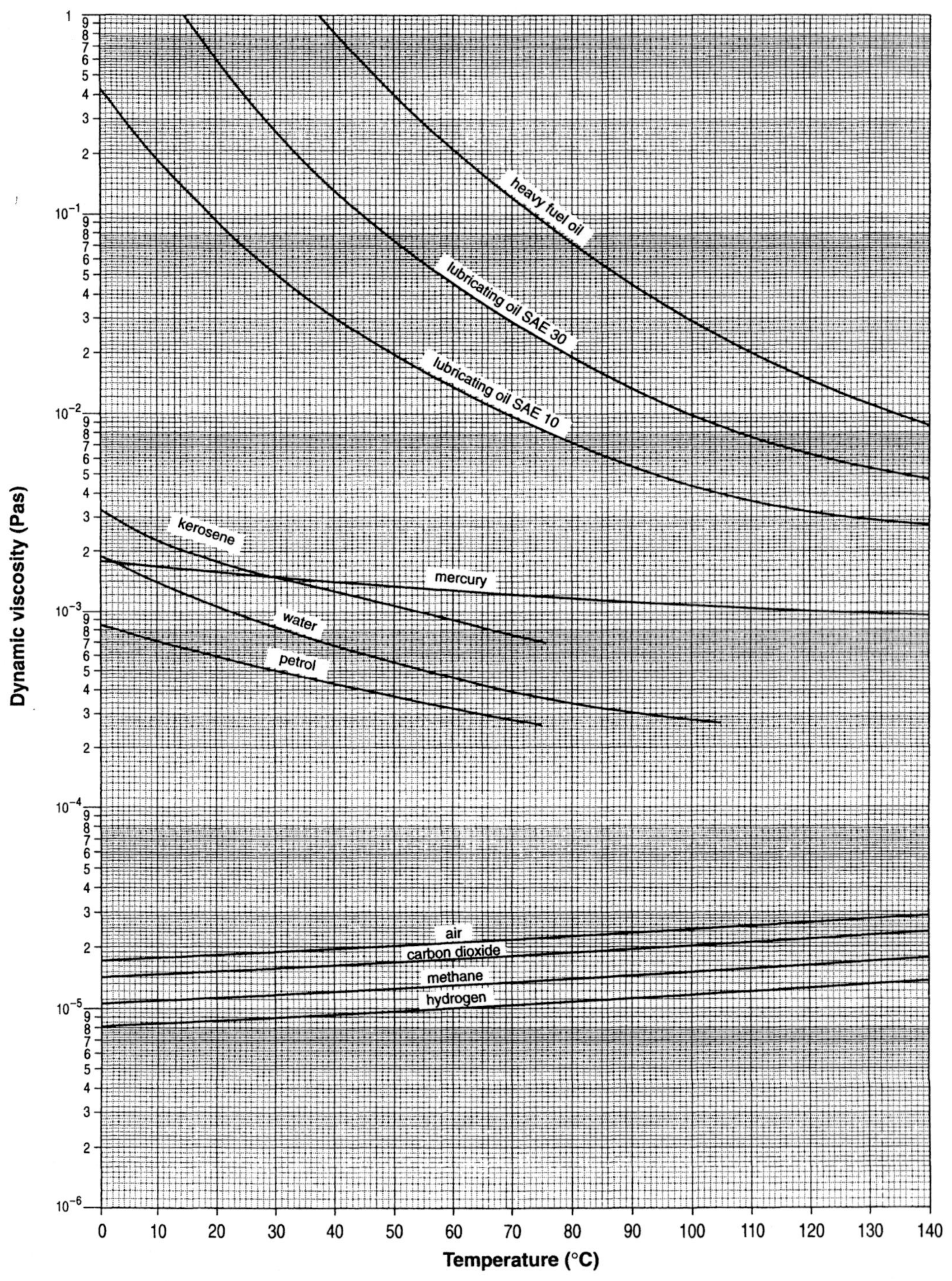

Appendix 4 Absolute roughness of various common pipe materials

Absolute roughness	mm
Cast iron	0.25
Commercial steel or wrought iron	0.045
Galvanised iron or steel	0.15
Concrete (cast on steel forms)	0.2
Concrete (spun)	0.1
Drawn tube	0.0015
Extruded tube made of metal, glass or plastic	0 (smooth)

Note The roughness values quoted are for pipes in the as-manufactured condition. The values are likely to increase with time, owing to the effects of corrosion, erosion and fouling.

Appendix 5   Moody diagram

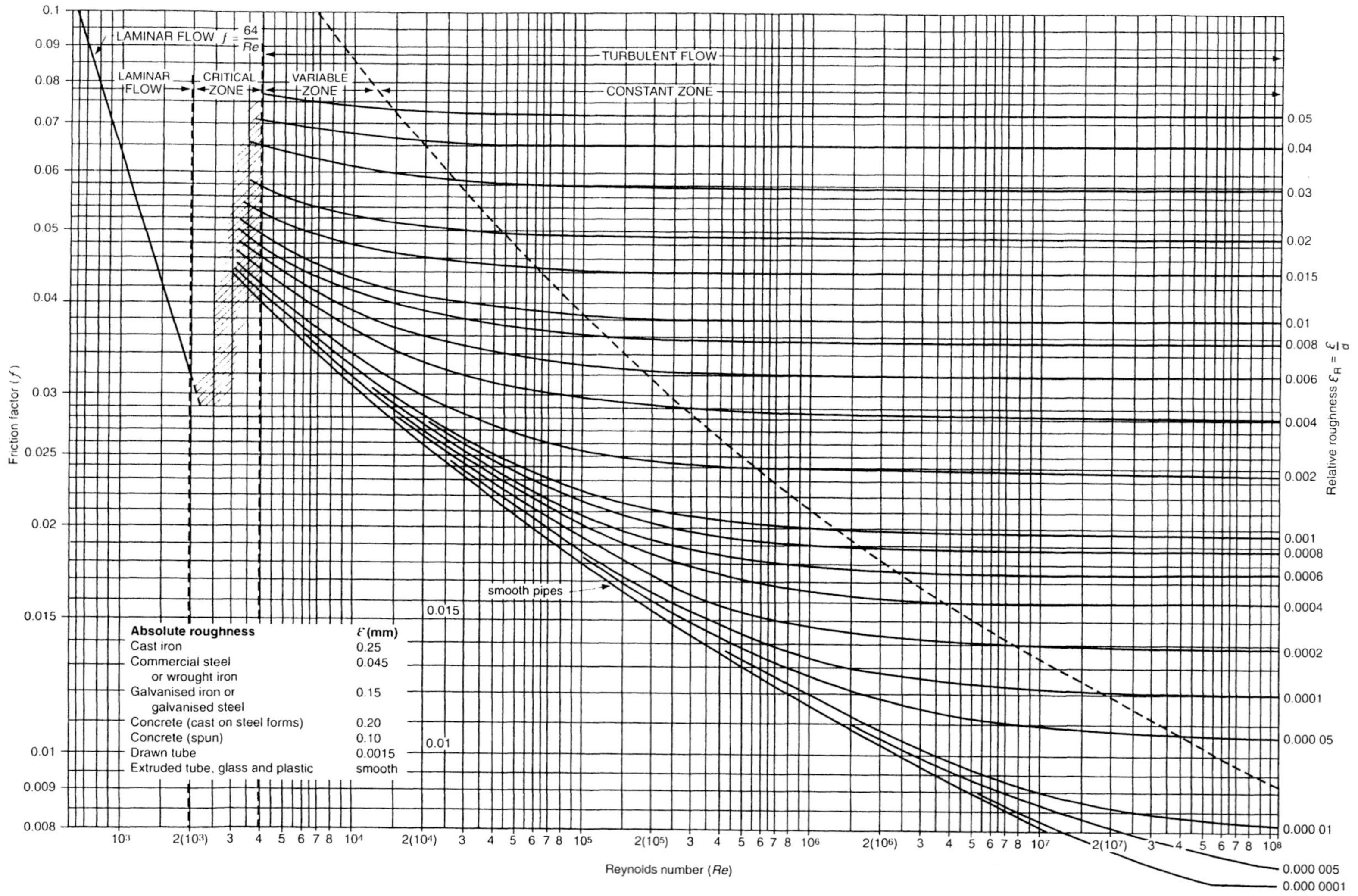

Appendix 6   *Simplified table of K factors for common fittings*

Fitting	K factor
45° elbow (standard radius)	0.3
90° elbow: standard radius	0.6
long radius	0.3
Return bend	0.8
Socket or coupler (screwed)	0.03
Tee: along line of flow	0.3
through side	0.8
Gate valve (fully open)	0.2
Globe valve: fully open	6.0
3/4 open	8.0
1/2 open	12.0
1/4 open	24.0
Check valve: hinged or swing disc	1.7
ball or poppet type	4.0
Foot valve with strainer:	
hinged or swing disc	3.0
ball or poppet type	7.0
Gradual transition: contracting	0 (negligible)
enlarging	0.75
Sudden contraction in a pipe	0.25
Sudden enlargement in a pipe	1.0
Sudden entrance (from tank to pipe)	0.5
Sudden exit (from pipe to tank)	1.0

Appendix 7   Series pipes

Refer to Figure App. 7.1 with two pipes A and B in series and their equivalent pipe E.

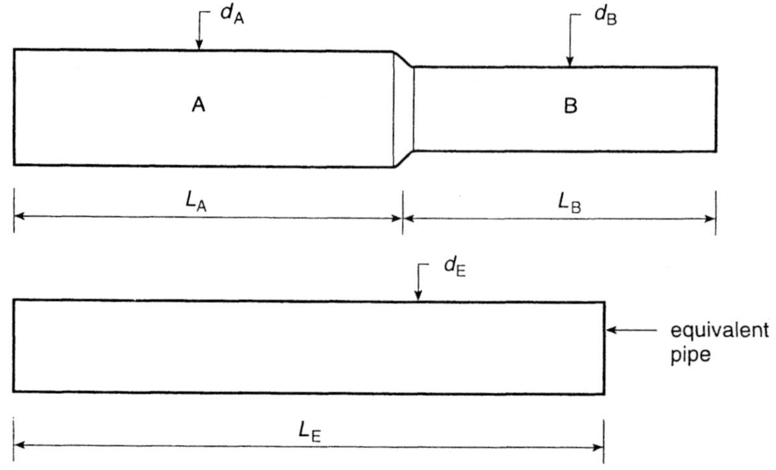

Fig. App. 7.1

$$H_L = f_A \frac{L_A}{d_A} \frac{v_A^2}{2g} + f_B \frac{L_B}{d_B} \frac{v_B^2}{2g} = f_E \frac{L_E}{d_E} \frac{v_E^2}{2g}$$

$$\therefore f_A \frac{L_A}{d_A} v_A^2 + f_B \frac{L_B}{d_B} v_B^2 = f_E \frac{L_E}{d_E} v_E^2 \qquad \text{.......................... (1)}$$

By continuity, assuming constant density of the fluid:

$$v_A d_A^2 = v_B d_B^2 = v_E d_E^2$$

$$\therefore v_A = \frac{v_E d_E^2}{d_A^2} \quad \text{and} \quad v_B = \frac{v_E d_E^2}{d_B^2}$$

Substituting in (1):

$$f_A \frac{L_A}{d_A} \frac{v_E^2 d_E^4}{d_A^4} + f_B \frac{L_B}{d_B} \frac{v_E^2 d_E^4}{d_B^4} = f_E \frac{L_E}{d_E} v_E^2$$

$$\therefore f_A \frac{L_A}{d_A} \frac{d_E^4}{d_A^4} + f_B \frac{L_B}{d_B} \frac{d_E^4}{d_B^4} = f_E \frac{L_E}{d_E}$$

$$\therefore d_E^4 \left[f_A \frac{L_A}{d_A^5} + f_B \frac{L_B}{d_B^5} \right] = f_E \frac{L_E}{d_E}$$

$$\therefore f_E \frac{L_E}{d_E^5} = f_A \frac{L_A}{d_A^5} + f_B \frac{L_B}{d_B^5}$$

Appendix 8   Parallel pipes

Refer to Figure App. 8.1 with two pipes A and B in parallel and their equivalent pipe E.

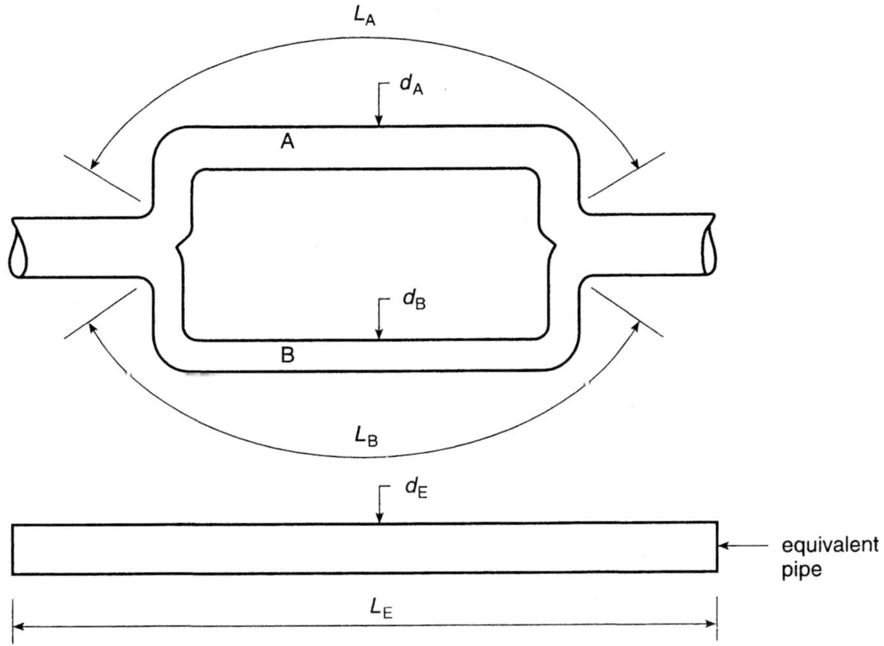

App. 8.1

$$H_L = f_A \frac{L_A}{d_A} \frac{v_A^2}{2g} = f_B \frac{L_B}{d_B} \frac{v_B^2}{2g} = f_E \frac{L_E}{d_E} \frac{v_E^2}{2g}$$

$$\therefore f_A \frac{L_A}{d_A} v_A^2 = f_B \frac{L_B}{d_B} v_B^2 = f_E \frac{L_E}{d_E} v_E^2$$

$$\therefore v_A^2 = \frac{f_E}{f_A} \frac{L_E}{L_A} \frac{d_A}{d_E} v_E^2$$

$$\therefore v_A = \left[\frac{f_E}{f_A} \frac{L_E}{L_A} \frac{d_A}{d_E} \right]^{\frac{1}{2}} v_E$$

Similarly: $v_B = \left[\dfrac{f_E}{f_B} \dfrac{L_E}{L_B} \dfrac{d_B}{d_E} \right]^{\frac{1}{2}} v_E$

By continuity, assuming constant density of the fluid:

$$v_A d_A^2 + v_B d_B^2 = v_E d_E^2$$

Substituting for v_A and v_B:

$$\left[\frac{f_E}{f_A}\frac{L_E}{L_A}\frac{d_A}{d_E}\right]^{\frac{1}{5}} v_E d_A^2 + \left[\frac{f_E}{f_B}\frac{L_E}{L_B}\frac{d_B}{d_E}\right]^{\frac{1}{5}} v_E d_B^2 = v_E d_E^2$$

$$\therefore \left[\frac{f_E}{f_A}\frac{L_E}{L_A}\frac{d_A}{d_E}\right]^{\frac{1}{5}} d_A^2 + \left[\frac{f_E}{f_B}\frac{L_E}{L_B}\frac{d_B}{d_E}\right]^{\frac{1}{5}} d_B^2 = d_E^2$$

$$\therefore \frac{1}{d_E^{\frac{1}{5}}} \left\{ \left[\frac{f_E}{f_A}\frac{L_E}{L_A}\right]^{\frac{1}{5}} d_A^{\frac{5}{2}} + \left[\frac{f_E}{f_B}\frac{L_E}{L_B}\right]^{\frac{1}{5}} d_B^{\frac{5}{2}} \right\} = d_E^2$$

$$\therefore \left[\frac{f_E}{f_A}\frac{L_E}{L_A}\right]^{\frac{1}{5}} d_A^{\frac{5}{2}} + \left[\frac{f_E}{f_B}\frac{L_E}{L_B}\right]^{\frac{1}{5}} d_B^{\frac{5}{2}} = d_E^{\frac{5}{2}}$$

$$\therefore (f_E L_E)^{\frac{1}{5}} \left\{ \left[\frac{d_A^5}{f_A L_A}\right]^{\frac{1}{5}} + \left[\frac{d_B^5}{f_B L_B}\right]^{\frac{1}{5}} \right\} = \left(d_E^5\right)^{\frac{1}{5}}$$

$$\therefore \left[\frac{d_E^5}{f_E L_E}\right]^{\frac{1}{5}} = \left[\frac{d_A^5}{f_A L_A}\right]^{\frac{1}{5}} + \left[\frac{d_B^5}{f_B L_B}\right]^{\frac{1}{5}}$$

Appendix 9   Manning coefficient

Typical average values of the Manning coefficient n with water flowing in large channels or open pipes.

Material	n $(\text{m}^{-1/3}\,\text{s})$
Smooth cement or ceramic	0.012
Cast iron (as cast)	0.015
Brickwork and masonry	0.018
Firm gravel	0.020
Corrugated pipe	0.023
Natural earth	0.025
Natural earth with large stones	0.032

Note   The above values of the coefficient assume a relatively clean surface. In time, pitting, slime or other marine growths may appear, and these will cause an increase in the value of the coefficient.

Appendix 10 Performance of a gear pump running at constant speed (1450 rpm)

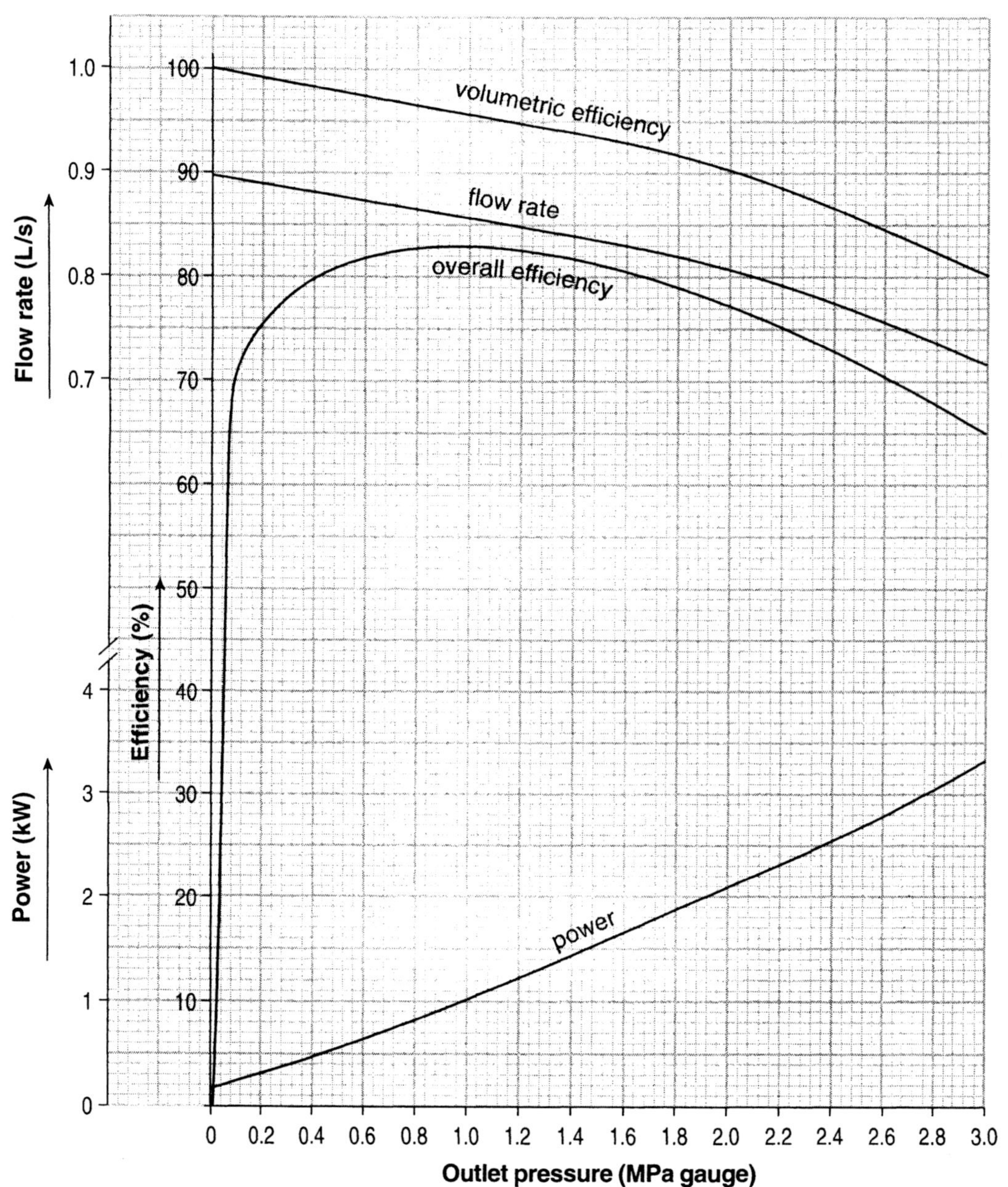

Appendix 11   Performance of a centrifugal pump running at constant speed (1450 rpm)

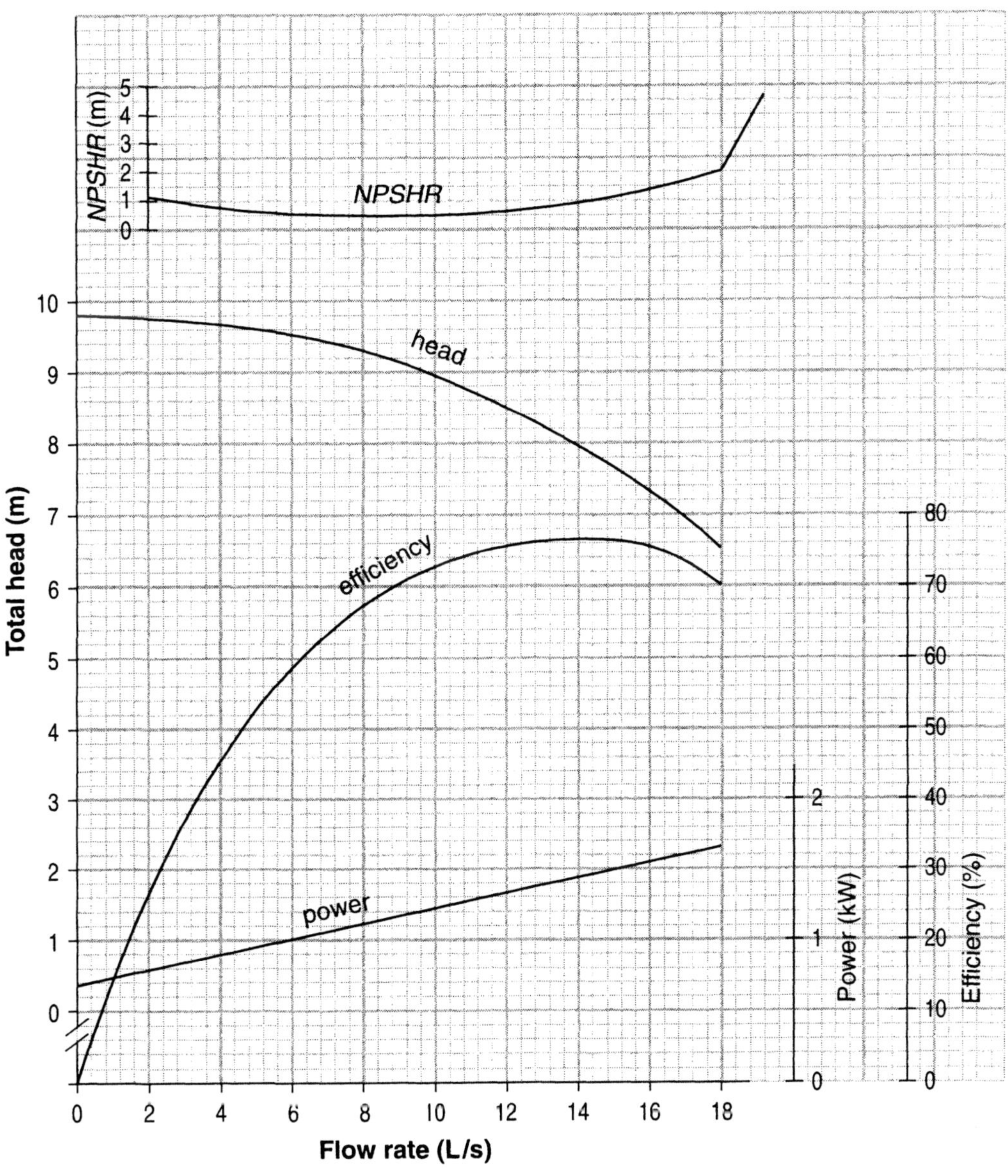

Appendix 12 Performance of a centrifugal pump with various impeller diameters at constant speed (1450 rpm)

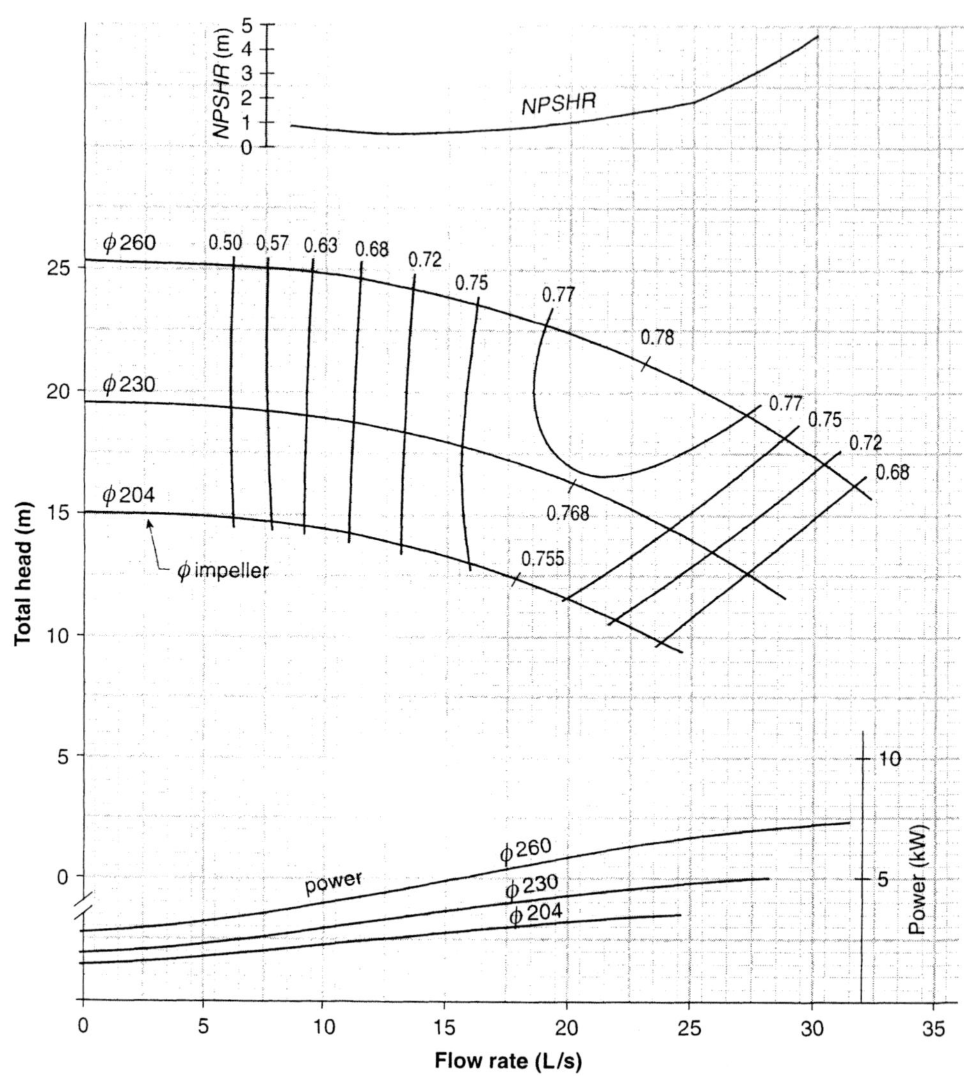

Appendix 13 Density, viscosity and saturation vapour pressure of water

Temperature (°C)	Density (kg/m^3)	Dynamic viscosity (Pas)	Vapour pressure (kPa (abs.))
0	1000	1.80×10^{-3}	0.61
5	1000	1.52×10^{-3}	0.87
10	1000	1.31×10^{-3}	1.23
15	999	1.15×10^{-3}	1.71
20	998	1.00×10^{-3}	2.34
25	997	0.90×10^{-3}	3.18
30	996	0.80×10^{-3}	4.25
35	994	0.72×10^{-3}	5.64
40	992	0.66×10^{-3}	7.38
45	990	0.60×10^{-3}	9.60
50	988	0.55×10^{-3}	12.3
55	986	0.51×10^{-3}	15.7
60	983	0.47×10^{-3}	20.0
65	980	0.44×10^{-3}	25.0
70	977	0.41×10^{-3}	31.2
75	974	0.38×10^{-3}	38.6
80	971	0.36×10^{-3}	47.4
85	968	0.34×10^{-3}	57.8
90	965	0.32×10^{-3}	70.1
95	962	0.30×10^{-3}	84.6
100	958	0.28×10^{-3}	101.3

Appendix 14   Range chart for centrifugal pumps

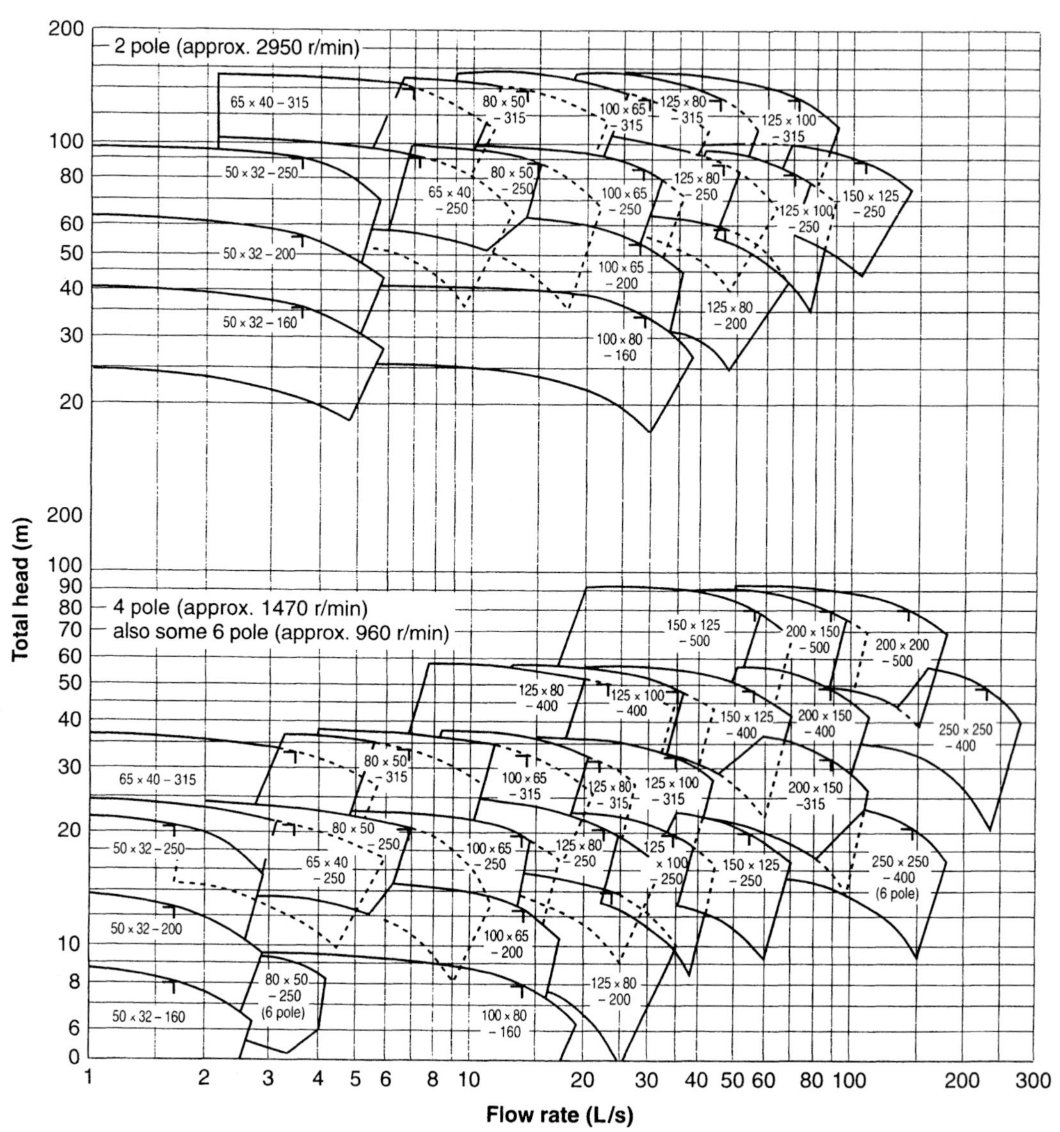

Index